RESERVOIRS

Also from the Macmillan Press

Engineering Hydrology, second edition
E. M. Wilson

RESERVOIRS

Brian Henderson-Sellers

Department of Civil Engineering
University of Salford

M

First published 1979 by
THE MACMILLAN PRESS LTD
London and Basingstoke
Associated companies in Delhi Dublin
Hong Kong Johannesburg Lagos Melbourne
New York Singapore and Tokyo

Typeset in 10/11 IBM Press Roman by
Reproduction Drawings Ltd, Sutton, Surrey

Printed in Great Britain by
Unwin Brothers Limited, Gresham Press
Old Woking, Surrey.

British Library Cataloguing in Publication Data

Henderson-Sellers, Brian
   Reservoirs.
   1. Reservoirs
   I. Title
   627'.86   TD395   470266996

   ISBN 0-333-24660-8

For Ann

There is nothing in nature more enchanting than a lake. Rivers I have loved, and with them the restless sea, so magical and yet so melancholy, perhaps because it seems the symbol of our desires; but it is those lovely lapping sheets of water, neither seas nor rivers, yet having the charm of both with something added, some touch of quiet, peace, soul's ease, that really possess my heart. You travel over leagues of hulking and stubborn land, then suddenly turn a corner and find a space where there is no earth, but only a delicate mirroring of the sky and that faintest rise and fall of waters, the lap-lap-lap along the little curving shore. Where else can you find such exquisite beauty and tranquility? May I end my days by a lake, one of earth's little windows, where blue daylight and cloud and setting suns and stars go drifting by to the tiny tune of the water.

J. B. Priestley, Essays of Five Decades (Heinemann, 1969)

Contents

CONTENTS

Preface

This book is intended to describe what a reservoir is, how it operates and why
indeed it is necessary at all. The subject is approached from many angles: engineering, biology, chemistry, management, ecology and mathematics. It thus should be
of interest to anyone wishing to gain a foothold in this interdisciplinary subject,
by learning a little about the views taken by specialists in fields different from,
but nevertheless closely allied to, his own interests. To facilitate understanding
by a wide readership, technical language and complicated mathematics have been
avoided. Some numerical examples are included. It is considered that, although
fuller understanding can be best gained by working through examples, the flow
of the text would be interrupted if too large a number were included. The reader
is therefore advised to consult the appropriate (single-subject) text and work
through examples on his or her own. For this (and many other) reasons a fairly
comprehensive bibliography is included, to direct the reader to texts that deal in
detail with the more orthodox approaches to this subject of reservoirs—a topic
which, as water demands continue to increase, must enter *all* our lives at home
and at work with increasing frequency and importance.

When units occur the SI (*Système International*) units will be used. When
older units are, perhaps, more meaningful or familiar these are included in
brackets.

Because natural lakes are often used as water-supply reservoirs (for example,
Haweswater, Thirlmere)—in effect a reservoir is but a man-made lake—the terms
'lake' and 'reservoir' may be regarded as interchangeable for our present purposes,
and will be used as such.

BRIAN HENDERSON-SELLERS

Acknowledgements

I am grateful to all those who have given assistance and taken an interest in the preparation of this manuscript; I especially wish to thank Professor E. M. Wilson and Dr S. I. Heaney for all their help during this project. I would also like to thank the many people who have had the time and patience to read the draft for this book and have proffered valuable advice. Of these, I would particularly like to express thanks to Dr M. W. Rogers and Dr J. F. Talling. In doing this I accept total responsibility for all opinions expressed herein—any remaining errors must be my sole liability.

I am also grateful to Mrs S. E. Mather for her excellent artwork and to the following authors and publishers who have given me permission to include several of their tables and figures: *Area* and the Institute of British Geographers for figure 2.3; Edward Arnold and A. C. Twort for table 1.2; the Controller of Her Majesty's Stationery Office for figure 11.1; the Division Director, Southern Division, Yorkshire Water Authority for figure 6.1; Freshwater Biological Association for figures 10.8 and 10.9; Gleeson Civil Engineering Ltd for figure 2.4; Dr D. H. Mills for figure 7.1; W. B. Saunders Company and Professor R. G. Wetzel for figure 7.6; Severn-Trent Water Authority for figure 4.6; Keith Smith for figure 11.4; Thames Water—Vales Division for figure 8.4; Welsh Water Authority for the cover photograph; the World Health Organization for table 12.1; R. E. Youngman for figure 8.3; also J. B. Priestley and William Heinemann Ltd for the quotation appearing on p. vi.

I also wish to thank the Anglian Water Authority for allowing access to their land for the purpose of obtaining photographs for figures 4.7, 4.9 and 12.1 and the North West Water Authority for figures 1.1, 1.2 and 2.1.

Finally I wish to thank my wife and family for all the advice and inspiration they have given me; without their encouragement the book would never have been written.

1

Introduction: Storage of Water for Potable Supply

Water is essential for life. Every person requires 0.0025 m^3 (2.5 litres) of *fresh water* each day simply to maintain the biological processes.

It is within this context that we examine one of the major links in the chain which permits a plentiful and wholesome supply of water to be made readily available—the *storage reservoir*. A reservoir plays a vital role in balancing supply and demand. An increasing population and/or an increasing standard of living are usually directly linked to a desire for a larger, continuous supply of water. Even in a temperate climate, water (in the form of rain or in streams) is not always available to an individual. To maintain supplies when natural sources fail, water must be collected wherever and whenever there is an excess, and *stored*. The many ways of accomplishing this and their associated problems are explored in the following chapters. First, however, it is worth demonstrating how this simple idea of water storage in a reservoir arose and whether there are any feasible alternatives.

Small Communities and Water Supply

When people began to establish settled communities, they nearly always chose a location near to a source of water—usually a river, but sometimes a spring, well, oasis or lake. For a small population a river is capable of supplying fresh water for drinking, cooking and irrigation and at the same time removing wastes from the neighbourhood. If washing is practised downstream of the 'watering place', any added impurities would not interfere with the drinking supply but would be slowly destroyed by the natural purification processes in the river. These processes depend primarily on the presence of certain organisms (such as bacteria) together with a sufficient supply of dissolved oxygen (see chapter 8), plus *time*. As long as there is no one in close proximity downstream who requires clean water, this biological self-purification by the river is accomplished without hazard to health. With a simple water system such as this, there is often little necessity to pretreat drinking water or to spend time and energy on purifying wastes.

Once the settlement has become well established, its water demand tends to

1

rise and may exceed the natural supply, which must therefore be artificially augmented. Water must be 'imported' (usually by pipeline) from a neighbouring area.

Although the problems incurred with treatment of waste water will not be discussed (see the bibliography for texts on this subject) it is important to note that, as communities expand towards one another, one town's wastes quickly become at least a part of the supply for its downstream neighbour. If the distance between the towns is insufficient for natural processes to render the water-borne wastes innocuous during the time of their journey, then the health and safety of the second town may be in jeopardy unless sufficient precautions are taken. Treatment of the water (see chapter 12) is then necessary to eliminate water-borne diseases such as typhoid and poliomyelitis.

Large Communities and Water Supply

In a large conurbation where the local demand exceeds the local supply, it is not only necessary to pipe water from another locale, but also to ensure that this second supply is dependable. This involves the use of two different types of reservoir: a storage reservoir and one or more service reservoirs.

A storage reservoir is sited where the supply is greater than the local demand. Water accumulates and is stored in a valley or large depression (either natural or man-made). The storage capacity is large enough to provide the total water required by the controlling community for several months—throughout the severest drought. This is the usual answer to today's water problems. Reservoirs have been built and used for many thousands of years, although their proliferation, both in number and size, is a more recent phenomenon (see chapter 2). The largest man-made lake in Great Britain, Rutland Water, was completed in 1976 and has been full since early 1977, supplying water to the Peterborough and Northampton area and providing recreational facilities for thousands of visitors. The changing role of these man-made lakes, becoming leisure centres as well as supplying potable water, is a recent phenomenon.

Suitable locations for storage reservoirs tend to be far removed from the centres of population that they supply. In order to balance the day-to-day demands for water, *service reservoirs* are needed within the city boundary. These are much smaller and contain water that has already been treated to a high degree. To maintain this water purity, they are covered and often hidden from view (see figure 1.1). The task of a service reservoir is to supply sufficient 'head' (that is, pressure) to maintain supplies to all houses in the vicinity (including multistorey blocks of flats) throughout a 24-hour period. When demand is highest (during the day, especially near meal times) the service reservoir becomes depleted. However, since it receives a continuous and steady supply of water from the storage reservoir (via the water-treatment works), it is refilled during the night when the needs of the town are lower. If the urban area is too flat, then these service reservoirs are often constructed in the form of *water towers* (see figure 1.2).

To determine the capacity needed for both service and storage reservoirs, some measure is needed of the water used by the community. This is usually in terms of the water *consumption*.

Figure 1.1 *Service reservoir at Croft, Cheshire*

Figure 1.2 *Water tower in Newton-le-Willows, Merseyside*

Water Consumption

In many cases the word consumption implies using a material so that it undergoes a chemical change (such as combustion). The chemical structure of the waste products will usually bear no resemblance to the original material.

When water has been 'consumed', the product is simply dirty water. It is possible to remove the wastes from this, resulting in relatively clean water. *Water consumption* therefore implies a temporary removal of water from its natural (*hydrological*) cycle. This is the cycle in which rainwater flows down the hills in streams and rivers to the sea. Evaporation permits water to enter the atmosphere, condense as clouds and eventually bring rain, thus completing the cycle (see figure 1.3).

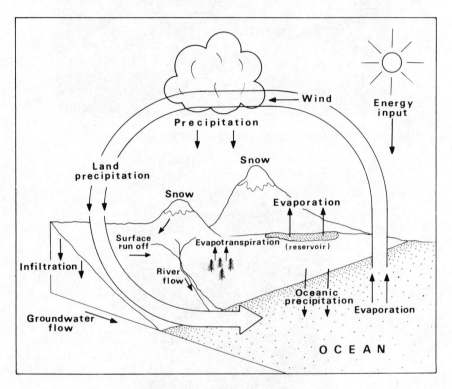

Figure 1.3 *The hydrological cycle*

If the total amount of water used by a population is divided by the number of people responsible for its use then we can derive some idea of the amount of water consumed, on average, by each person (that is, per head). This is known as the *per capita consumption*. If this calculation is done for the total amount of water used in the home then the *domestic per capita consumption* results. This figure takes into account drinking, cooking, washing, laundry, gardening and waste disposal. Figure 1.4 shows how this amount is divided between the different usages, based on U.K. figures for 1976.

These recent figures, derived from simultaneous studies in Mansfield and Malvern, show that earlier values for consumption obtained by cruder methods were overestimates. The typical 1970 figures quoted are usually approximately 0.140 m^3/head/day, excluding waste in distribution. The recent figures show that domestic consumption appears to be about 30 per cent less than previously calculated. Since the total demand of an area can be measured at the source (for example, the reservoir) it seems that there is a much larger waste in distribution (possibly by as much as a factor of 3) than earlier figures suggest. (Although the 1976 figures given in figure 1.4 were obtained during a drought period, they have been substantiated by the corresponding consumption for 1977.)

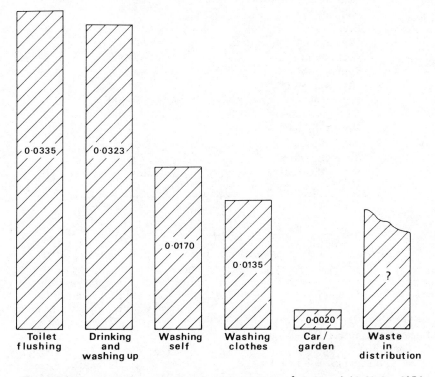

Figure 1.4 *Average daily per capita water consumption, m³, in Mansfield, Notts., 1976 (adapted from Thackray, Cocker and Archibald, 1978)*

In addition to water used for drinking, cooking and washing, an industrialised society makes further demands on a water supply network. Many industries require vast quantities of water in order to produce their manufactured products (see table 1.1). A large percentage of this is used for washing and cooling and thus may be of low quality; approximately 40 per cent of the water used by the Central Electricity Generating Board for cooling is saline or brackish. On the other hand, some industries, notably food processing, require a very high water quality.

The total amount of water used by industry can be considered in isolation. However it is more usual to allocate a proportion of this to each individual within the country. If the total amount of water used by industry is divided by the total population then this gives some measure of the industrial consumption per head of the population. If this *per capita* consumption is added to the domestic *per capita* figure, then the *total per capita consumption* is found. It is this figure that

Table 1.1   Industrial Consumption

Manufacturing process	Water used (m³/tonne of product)
Acetic acid	450
Ammonium sulphate	900
Baking	4
Beer	10–20
Caustic soda	90
Coal mining	2–5
Laundering	45
Milk processing	4
Paper making	30–300
Steel production	5–100
Sugar refining	8
Sulphuric acid	4–25
Synthetic fibres	140
Synthetic rubber	2500
Vegetable canning	10

Table 1.2   Total *per capita* Demands for Various Regions and Cities (reproduced, by permission, from Twort *et al.*, 1974)

Country	*Per capita* demand (m³/day)	litre/day
England and Wales (1971–2 figures)	0.290	290
Scotland (1971–2 figures)	0.415	415
Other countries (1967–8 figures)		
Denmark	0.340	340
France	0.300	300
U.S.A.	0.250–0.350	250–350
Sweden	0.210	210
West Malaysia	0.164	164
USSR municipal	0.164	164
Johannesburg native housing	0.118	118
Native villages where water is hand carried	0.015–0.035	15–35

is used to decide how much water is likely to be needed in the future. This is the anticipated *demand* and will be equal to the consumption only if the estimates of the Water Authority are 100 per cent accurate.

Table 1.2 gives figures for the total *per capita* consumption for several different countries in various parts of the world, illustrating how water consumption is related to the degree of 'development'.

Sources of Water

Rivers

The most direct and readily available source of water is a river. A river is the main
artery which transfers water from the catchment areas in the hills to the sea. It is
seldom sufficient in itself as a large-scale dependable supply since the volume of
water carried in it fluctuates rapidly. However it is a prime source from which
water can be abstracted at any point on its course and put into storage. A river
may feed a reservoir directly or water may be abstracted from the river and
pumped overland to the reservoir. When rivers are used in this way, care must be
taken to ensure that there are no harmful effects to downstream river users.

Rivers often transport effluent to the sea. In both the river and the sea, bacteria degrade the organic wastes in the water and purify it, providing there is
sufficient dissolved oxygen present. To provide enough oxygen for this process
a minimum volume of water must be present in the river *at all times*. The lowest
volume of water needed for this is known as the *minimum acceptable flow*
(M.A.F.) or, alternatively, as the *compensation water*. (This may be greater or
less than the *dry weather flow*, D.W.F.–an alternative method of describing
periods of low flow, which as yet has no unique definition.) Failure to maintain
the M.A.F. may result in anaerobic conditions arising in the river. The end product
of degradation under anaerobic conditions (this time by types of bacteria that do
not need dissolved gaseous oxygen) include hydrogen sulphide and methane
(marsh gas), which are undesirable and unpleasant.

Alternative Sources

An alternative natural source of water is found in wells and springs. Both these
sources are related to the *groundwater*. When rain falls on the land it may flow
straight into a river (a process known as *runoff*) and so becomes quickly available
as a water supply. Depending upon the state of the ground (that is, how much
water it already contains) a proportion of this precipitation may percolate down
into the rocks. Some rocks are more permeable to water than others. Rock strata
that will hold water are known as *aquifers*. An aquifer is usually bounded on the
bottom (and sometimes on the top) by an impermeable stratum, so that the
water in an aquifer is only available for use by borehole, from a well or when the
aquifer reaches the surface (usually on a hillside) to form a *spring*. These cases are
illustrated in figure 1.5. As a natural water source, groundwater has been in use
for centuries. More recently (the last hundred years or so) groundwater has also
been added to the list of artificially maintained sources. Water is removed from
the aquifer by pumping–a natural or artesian well needs no pumping when the
head of water is sufficient (see figure 1.5). As the level of groundwater (often
referred to as the *water table*) falls, more pumping is necessary. During times of
heavy rain the aquifer will *recharge* itself naturally, although this may not be fast
enough to satisfy the pumping rate indefinitely. In this case it is necessary (and
economically viable) to pump excess water *back* into the aquifer to replenish it.
Such a scheme to recharge the chalk aquifer under London was begun in 1977.
If successful, it may mean that the fountains in Trafalgar Square, originally
powered by the natural water pressure in the ground but now pumped, may once

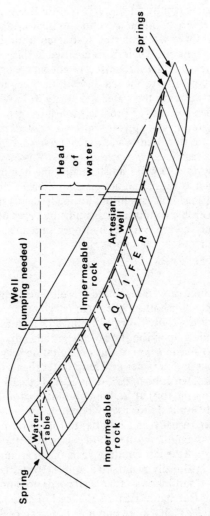

Figure 1.5 *Geological structure of water-bearing rock strata (aquifers), illustrating the difference between springs, wells and artesian wells*

again flow without direct intervention.

The use of groundwater from aquifers is limited by the number of suitable strata that are available. Aquifers represent underground storage and as such can be regarded as subterranean reservoirs. They occupy no land surface and so no valuable agricultural or building land is lost. The water contained in them is purer since it is less open to atmospheric pollution. However this fact in itself may sometimes be a severe handicap. Unlike rivers, where the passage of water from the hills to the sea can be measured in days or weeks, the time taken by groundwater for the same journey can be years or centuries. *If* any pollutant is accidentally introduced into the groundwater, then its removal by natural flushing is a very long procedure.

Groundwater schemes are at present being implemented, often within a supply network that includes aquifers, surface reservoirs and rivers. These supply systems are referred to as *conjunctive schemes*. Restricted by the natural geology, aquifers may have only a limited role in the development of water resources.

Surface Reservoirs

A surface reservoir is the means most often used for collecting and storing excess water, which is then released at times of the year when demands exceed the supply. Unlike many gases, which can be compressed and stored in small cylinders, water is virtually incompressible and therefore its storage utilises a vast acreage. Large tracts of valuable agricultural land may be lost—a consideration that ranks high on the topics for debate when a new reservoir is proposed. Two basic types of storage reservoir can be built: an *impounding* reservoir and a *pumped storage* reservoir.

The concept of an impounding reservoir is the older of the two. A stream is dammed and the water upstream of the dam collects in a lake behind the dam. This is a good solution if the volume of water in the stream is sufficient to fill the lake fairly quickly. Streams and valleys where this technique can be usefully employed abound in mountainous regions of high rainfall, such as the Snowy Mountains of Australia and the Welsh and Scottish Highlands in Great Britain. Such reservoirs are often far removed from centres of population and demand, and so transportation costs must be included in the evaluation of new schemes.

If the reservoir is to be sited in a valley where the flow of the incumbent stream is too low to fill the reservoir within an acceptable time (a few years at most), then the scheme must be of the pumped-storage kind. In this, the natural inflow of a stream is replaced or supplemented by an artificial pumped inflow. The water is abstracted from a nearby large river when it is in spate and pumped into the reservoir for storage.

Two modes of operation exist for any reservoir: the *direct supply mode* and the *regulating mode*. In the former case, water is taken from the reservoir when it is needed, sent to a treatment works for purification and thence directly into the water mains network. To use the reservoir for river regulation, water is released at times of low flow into the effluent stream. This must be done to maintain the M.A.F. Furthermore, if the flow in the river can be kept at a constant value greater than the M.A.F., then this extra volume can be abstracted for use further downstream. Transportation costs are thus reduced, since the abstraction point can be arbitrarily near the demand centre. Most reservoirs are able to operate in either of these modes, although strictly regulating reservoirs do exist (for example, Kielder). Further uses of reservoirs will be discussed later; these include schemes for providing hydro-electric power and flood-alleviation projects (see chapters 2 and 11, respectively).

Summary

Because it is necessary to maintain a disease-free supply of water to a community, techniques of collection, storage and treatment of water are important to both industrialised and developing nations. Efficient water storage is vital. Collection is only useful if the collected water can be stored; and treatment and supply are only possible, on a large scale, if stored water is available. The storage of water in reservoirs is crucial in most water supply systems around the world not only for direct human and animal consumption but also for irrigation, industrial processing and hydro-electric power. (The New High Dam at Aswan was built to supply electricity and not water to the Nile Valley!)

The measurement of water usage is expressed in terms of water consumption, thus providing a quantification for both present use and future demands. Although rivers, wells and aquifers are being used, the major factor in most new supply schemes is still the surface storage reservoir. Recent developments in pumped storage schemes (for example, Rutland Water) have led to larger and larger artificial lakes, while socio-economic and environmental pressures demand their secondary use as recreational centres.

Suggested Reading

Smith, K., *Water in Britain* (Macmillan, London, 1972).
Twort, A. C., Hoather, R. C., and Law, F. M., *Water Supply*, 2nd ed. (Edward Arnold, London, 1974).

2

The History of Water Supply

In their history, freshwater reservoirs are not, as one might initially assume, inextricably linked with water supply. Although the construction of dams and barrages to impound a body of freshwater dates back almost 5000 years, this stored water was used primarily for irrigation and as a source of providing water for canal transportation. Only during the time of the Roman Empire and for the last 150 years in modern Europe and America have artificial lakes been used as a water supply.

Dams in History

The oldest known dam appears to be that at Sadd el-Kafara across the Wadi el-Garawi. The full history of the dam is unknown and little of it remains today. It was built, arguably, in about 2800 B.C. to supply water both for drinking and to assist in stone extraction at the nearby alabaster quarries. But the workmanship was inappropriate; no mortar was used.[1] The dam was thus not watertight and consequently lasted only a short time. The Egyptians had little use for irrigation reservoirs, since the yearly flooding by the Nile provided an efficient system of basin cultivation, and the art of dam building was never developed within their culture.

About a thousand years later, the civilizations developing in the 'fertile crescent' between the Rivers Tigris and Euphrates built a complex system of diversion dams and irrigation canals for their agriculture. Many of these were small and temporary structures (often of earth and wood) and few were intended to create water storage in a reservoir.

A more unusual use for a reservoir was devised by Sennacherib in 689 B.C. In the campaign against the Babylonians, he put a dam across the Euphrates until a large lake had formed behind it. He then breached the dam, and the subsequent torrent of water destroyed Babylon. This was the first but not the last act of war in which a reservoir has figured; another example is the breaching of the Möhne

---

[1] The Egyptians did not use mortar as a sealing agent, but only as a lubricant. They relied on the weight of the structure itself to give stability, as in the case of the Great Pyramid.

and Eder dams in the Second World War.

The Romans pioneered the use of reservoirs for water supply, developing a technology that also permitted reservoirs to be used for irrigation, flood control and river navigation. The concept of a reservoir dam is one of their highest achievements. Indeed the first man-made recreational lake was created for Nero: the quality of the view from his villa was enhanced by a system of three reservoirs. Unfortunately it was probably this same sophisticated water-supply system that led to the decline of the Roman Empire: lead poisoning became widespread as a result of the use of lead water pipes and lead drinking vessels. With the fall of the Roman Empire, the art of reservoir storage became lost for many centuries.

Water for Transport

In Europe, irrigation has never been practised on a large scale. The first need for reservoirs was to provide water for 'flash locks' for the river navigations that were beginning to be constructed. The earliest reservoir in Great Britain was built in 1189 near Winchester, to serve the recently constructed River Itchen navigation. The Exeter canal (built in the sixteenth century) also used such a system, and an early dam built by Smeaton on the River Coquet served a local ironworks. It was the advent of the Industrial Revolution, characterised (in this case) by canal transport that gave the greatest impetus to reservoir construction.

The Peak Forest canal was built between 1794 and 1800, requiring two reservoirs to provide sufficient water for the highest lengths. The larger of these two dams is 20 m high and over 200 m long. These and many others constructed for the canals in about that period set the design standard for many of the reservoir dams to follow.

Subsequently population and standards of living (reflected in both personal and industrial demands) have increased. These have been the major factors leading to the inadequacy of established sources of water (from springs, wells and rivers) and to the growth of technology for dam and reservoir construction to augment the water supply to major conurbations.

Water for Power

Over the same period, water power was used in many small industries (for example, corn grinding) although the water wheel was rapidly replaced by steam engines. In the last hundred years the demand for energy has resulted in an increase in electricity generation. Water turbines play a significant part in power production in many countries of the world. To generate this *hydroelectric power*, water is stored in a highly elevated location and then allowed to impinge vertically on to the turbine blades. The greater the vertical displacement, the more potential energy is available for conversion to kinetic energy and hence (by means of the turbines) to electrical energy. Hydroelectric schemes are therefore ideally suited to upland reservoirs situated on a plateau edge (for example, in Scotland, Norway, Sweden) or as an integral part of a large dam. In the latter scheme the generating station is situated at the foot of the dam itself and utilises the water released through the dam wall from near the top water level (T.W.L.), as at Aswan (Egypt) and Kariba (Zambia/Zimbabwe). Hydroelectricity may thus be an added bonus from (or sometimes the main reason for) the creation of large reservoirs and may ensure that the project is both economically possible and environmentally worth while.

Water Supply

From the time of the Roman Empire until the beginning of the nineteenth century, little use had been made of reservoirs for public water supply. In Britain, the earliest example of a water supply reservoir appears to have been one at Whinhill constructed in 1796 to supply water to Greenock (Scotland). Initially it met with only limited success, failing structurally three times in its first forty years of existence.

At that time, in the reign of George III, the population of Great Britain was only $10\frac{1}{2}$ million. They had very little piped water and no form of sewerage system. Today a population over five times greater demands and enjoys the benefits of both. After its slow birth in the early nineteenth century, the water-supply industry has grown rapidly in the present century. Scotland did much of the pioneering work in this field; an artificially augmented supply to Edinburgh was the work of three famous engineers: John Smeaton, Thomas Telford and John Rennie. The resulting Glencorse Dam was completed in 1823 and was copied five times in the following three decades.

In 1848 the first major system of water supply reservoirs in England was begun in the Longdendale Valley, to supply Manchester (see figure 2.1). Already, prior to 1840, more than a dozen dams (of the order of 20 m high) had been constructed, almost all of them in the northern part of the country. For example, a dam on the River Rivelin was completed in 1836 to serve the city of Sheffield, and earth-

Figure 2.1 *Longdendale Reservoirs–serving South Manchester*

works were built on a larger scale by the Great and Little Bolton Waterworks Company, primarily to supply power for mill owners downstream; this system today forms part of the complex network of reservoirs feeding the Bolton area.

In France only one supply reservoir, the Zola Reservoir (named after its engineer), was built before 1850. The dam was made of rubble masonry and is still watertight today, constraining a lake 8×10^5 m^3 in volume. One of the earliest American dams, across the Schuylkill River at Fairmont, had a dual role, as public water supply and source of water for locks for navigation.

The succeeding years saw a steady growth of dam and reservoir technology, and, correspondingly, a steady increase in the amount of available water storage. Construction techniques are now based increasingly on theoretical considerations, in addition to practical experience. Fashions (and local geological requirements) have demanded earth dams, rock dams, masonry dams and dams of different shapes and orientations. Figure 2.2 shows the total number of dams over 16 m (50 ft) in height constructed in Great Britain between 1796 and 1966 (split into four periods) and also shows the ratios of earth to concrete structures. A recent survey of dams over 15 m high constructed during the period 1840–1971 indicates that the British situation is not mirrored on a larger scale. World dam construction

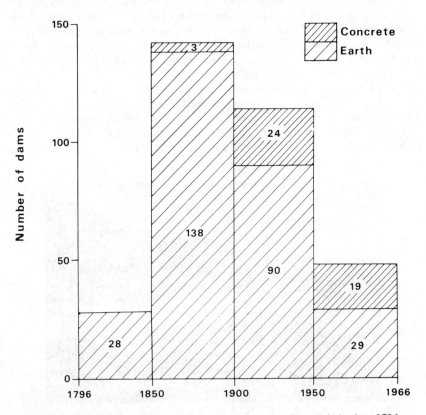

Figure 2.2 *Ratio of concrete: earth dams constructed in Great Britain since 1796*

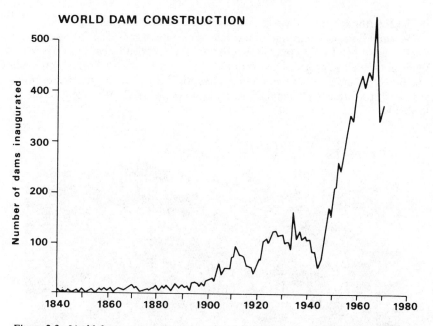

Figure 2.3 *World dam construction (reproduced, by permission, from Beaumont, 1978)*

peaked, not in the nineteenth century, but in the period 1945–71 (see figure 2.3), when a total of 8180 major dams were built—548 in the peak year of 1968. Most of these dams are in the North American continent (especially the United States), where over 200 major dams were being commissioned each year during this period of feverish activity. In the last decade, however, the rate of dam construction has slowed, which may reflect the growing world-wide concern with the environmental (and economic) impact of such large construction projects.

Dam failures have been few. In 1864 the Dale Dyke Dam near Sheffield failed and 244 lives were lost, while in 1925 16 died when the concrete Eigiau Dam was undermined. As a result of these tragedies, stringent safety precautions were devised and an Act was passed in 1930 to ensure that all reservoirs larger than 22 700 m^3 (5 million gallons) are inspected every ten years. (This Act has been updated by the Reservoirs Act of 1975.) One of the worst disasters in more recent years was the collapse of the earth Teton Dam (United States) in June 1976, killing 14 people and causing $400 million worth of damage. This is the largest dam that has failed to date, although at a mere 95 m high it doesn't qualify as a 'large dam'.[1] In 1975 62 such dams were in operation in the world and 25 under construction; only 13 of these were built before 1960. The world's worst dam disaster was probably the failure of the 23 m high South Fork Dam (United States) in 1889, which killed between 2280 and 10 000 people. Almost equal devastation resulted from a flood wave 100 m high that topped the 206 m arch Vaiont Dam in Italy in 1963. This wave was produced by an enormous landslide of material into

---

[1] 'Large dams' are dams over 150 m high. The first large dam was the Boulder Dam, built in 1936.

the lake. Although the dam was not breached, 3000 lives were lost.

As water demands continue to increase, more sites are needed for reservoirs. The easiest type of valley to dam successfully is narrow and steep-sided (as in figure 2.4); such valleys usually occur near the source of a river (often in the hills). The older, upland reservoirs stand in contrast to many of the larger, newer reservoirs built (or presently under construction) in the wide, flatter valleys of the lowlands.

Figure 2.4 *Castlehill Dam, Glendevon, Dollar, Clackmannanshire (photograph by courtesy of Gleeson Civil Engineering Ltd)*

Upland and Lowland Reservoirs

The advantages of an upland reservoir are many. The construction of the dam is relatively easy and thus costs are low. In some countries (for example, Great Britain, Norway) highlands are near the coast that faces the prevailing, rain-bringing winds. Frontal rain is augmented by orographic rainfall and these regions of high altitude and high annual rainfall make ideal locations for water-collection and storage schemes. In addition, the rain collected is relatively pure. Little pollution is contained in the air mass due to the sea passage and often there is no industry present in the catchment area itself (usually mountainous terrain). In high, shaded valleys the temperature rise of the water during summer is restricted and few forms of life exist. The nutrient load is also low and the lakes are oligotrophic (see chapter 8). The quality of the water stored is thus good.

Two major drawbacks do exist with upland reservoirs. The first is that of transportation costs; the gathering grounds are far removed from the centres of industry and population, and thus from water demand. This cost is offset by the high quality of the water received, which requires little conventional purification treatment. Manchester's water (after minimal chemical purification) is piped from the Lake District. The second problem is that of unavailability—sufficiently large upland valleys are no longer available. In recent years, this has necessitated the introduction of a new concept: the lowland reservoir.

Nearer centres of population, river valleys tend to be more mature—broader and shallower—although the rivers themselves carry a far higher volume of water. Damming a major river at any point along its lower reaches would quickly produce a lake of stored water large enough for most purposes. However the high density of development in the British Isles means that all such sites coincide with a major centre of population and/or industry. The only lowland valleys large enough to store the required volumes of water without destroying an urban area are situated on minor tributaries. The recently completed Rutland Water is situated on the River Gwash—a stream that would fill the lake (maximum capacity 124×10^6 m^3) in something over 200 years. To utilise such sparsely populated valleys, the natural stream flow from the catchment is usually supplemented by pumping water from a nearby major river (probably in a different catchment area) overland into the *pumped storage* reservoir.

Lowland reservoirs thus contain water abstracted at a point well downstream, where pollution levels may be high. Nutrient loads may be sufficient for the pumped-storage reservoir to become rapidly eutrophic (see chapter 8) and silting may become a problem. The increase in productivity has been accelerated, for instance, in the recently completed Farmoor Reservoir in the Thames Valley; here, low water quality necessitates the introduction of complex and costly purification methods before the water becomes potable.

Even in the more sparsely populated lowland valleys, reservoir construction may mean that some inhabitants will lose their homes and livelihoods. Valuable agricultural land may be submerged and the total ecological structure of the area changed. During the filling of the Cow Green Reservoir in Teesdale, conservationists were concerned about the possible loss of the unique Teesdale violet and the rare spring gentian, which were to be found growing below the proposed top water level (T.W.L.). Monuments, churches or halls of architectural interest may be threatened with submergence. Sometimes these losses will be accepted philosophically; on other occasions large amounts will be spent to save them for posterity. During the construction of the New High Dam at Aswan, dozens of monuments in Nubia, including two temples to Rameses II over 20 m high, were hewn out of the rock faces at Abu Simbel and removed piece by piece, to be re-erected on a new site a few kilometres away on a plateau above the water level.

In future, possible sites for new reservoirs must be carefully considered. Rutland Water and Foremark are both lowland (pumped storage) reservoirs, whereas Kielder is to be impounding. Economics, spatial dwelling patterns, nutrient loads, water quality and technology are all interrelated and all need to be taken into account before a decision (and the ensuing necessary Parliamentary Act) can be made; each of these aspects will be discussed in later chapters.

Alternatives for the Future

Many future developments have been discussed already: more and larger surface storage reservoirs, groundwater schemes and conjunctive use. It is worth mentioning some of the possible alternatives, although detailed reference is impossible; further recommended reading will be found in the bibliography. One of the major alternatives currently being developed is desalination. This set of techniques (such as reverse osmosis and multistage flash distillation) is uneconomic in countries where rainfall is moderately high, since desalination is, in general, energy intensive. However, many countries with low rainfall, especially those in the Middle East, are at present finding that desalination is the optimum solution to their water supply problems. These countries are so desperate in their need to obtain large quantities of fresh water that experiments have been instigated to tow large icebergs from the Polar regions—a technique which, surprisingly perhaps, is technologically feasible. In some fertile areas there is already a long and intricate history of community water management schemes.

In countries where water is readily available but land is precious, storage can be undertaken by reclaiming coastal land and creating freshwater reservoirs in estuaries or as isolated impoundments just offshore. Hong Kong is at present undertaking large scale damming of sea inlets. 1973 saw the completion of the Plover Cove Reservoir; the High Island Reservoir, due for completion in 1979, utilises a narrow channel between High Island and the mainland peninsula of Sai Kung.

In Great Britain, barrages across the Wash, Morecambe Bay, the Dee Estuary and the Solway Firth have been proposed and scale modelling has been undertaken. A barrage across the Severn Estuary has been proposed, primarily to utilise the large tidal range for power generation. The water stored upstream of the barrage may also provide a source of low-quality water for industry. There are many problems, including disturbance to the marine ecosystem, alteration of coastal silt movement and the costs of purification and transportation. Water is abstracted from the hydrological cycle at its lowest (topographical) and thus most polluted location. However, in this country economic pressures have caused these projects to be shelved until at least the next century. In the immediate future more emphasis may be placed on conservation and re-use (see chapter 12).

Summary

The use of reservoirs for water supply is, historically, a recent phenomenon. Although the Romans stored water in such impoundments, the real growth in reservoir use has occurred, in parallel with growth in population and energy consumption, during the last hundred years. When water demands first began to outstrip existing natural supplies, technologies used in the construction of reservoirs for feeding canals were adopted for the rapidly expanding water-supply industry.

Most early dams and reservoirs were constructed in the hills—in Britain along the western edge of the country, in Wales, Scotland and the Lake District. More recently, new reservoirs have been sited in the lowlands, nearer to centres of demand but holding water of poorer quality. Although transportation costs are cut, the costs of treatment plus the possible added expense of 'importing' water by pumping it overland to fill the reservoir are high. Many of the same arguments apply also to proposals for estuarial barrages. The present economic climate makes their implementation in the near future unlikely.

Continuously increasing demands for water can be satisfied by utilising more and more land for storage. The compromise between obtaining more water and retaining sufficient agricultural land areas must be reached both locally and nationally. In the longer term there must be a maximum possible water consumption for each autonomous country (or region). Increases in industrial demand may well be moderated by advances in technology and the industrial contribution to total *per capita* demand may even be reduced. However, if demands do continue to increase, and approach the maximum available water in the country, then some tempering of our domestic and industrial water consumption and thus our standard of living will become necessary.

Suggested Reading

Smith, N., *A History of Dams* (Peter Davies, London, 1971).

3

Reservoir Size

Specifications for a new reservoir must include its location and size. Both these depend on where the extra demand is forecast and how large the reservoir yield must be in order to meet this demand. To minimise transportation costs, the locality of the reservoir will be defined by its proximity to the centre of demand. This may itself eliminate the possibility of either an upland or a lowland reservoir. The necessary reservoir size is calculated by extrapolation of the available data on water demands, population growth, etc.

Data and Extrapolation

Extrapolation is a graphical technique for forecasting future behaviour. If we assume that demand is increasing then a graph of demand against time might look like those in figure 3.1. Figure 3.1a shows a demand that is increasing linearly with time. For a supposed community of 10 000 inhabitants, at the end of each year, the daily demand of the town has increased by 4 m^3 (4000 litres). If this trend continues (dotted line) then we may predict the demand for the year 2000.

The second graph (figure 3.1b) shows a case in which the upward trend is less regular. If the trend of 1940–60 is followed (that is, if 1961–77 is a temporary effect), then forecasting is best done by extrapolating the curve by the dotted line (I). If, however, the trend over the last 17 years is indicative of the future then line II would give a better forecast. The difference between these two estimates for the year 2000 is 2100 m^3 (2.1×10^6 litres) per day. Even if it can be safely assumed that extrapolation of the last 17 years' data will be most successful it remains the case that, unless part of the curve is linear, the straight-line extrapolation itself (line II) must be subjective. This introduces possible errors and thus the forecast figure is usually qualified by assigning it a probability of occurrence. The range of values within which the correct answer will lie with a high degree of probability (95 or 99 per cent) gives the *confidence limits* for the extrapolated curve.

As we have seen, extrapolation of a linear curve (straight line) is easiest. It is often possible to adjust scales on a graph so that the curve is 'distorted' to look like a straight line. If this is possible then a forecast is more easily made. For

20

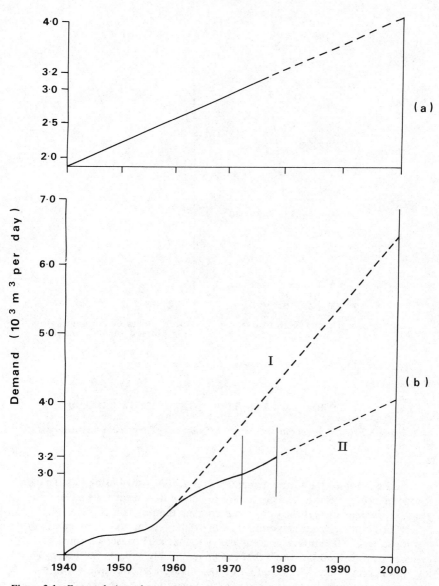

Figure 3.1 *Extrapolation of a trend: (a) a linear increase gives a useful forecast; (b) a non-linear variation makes prediction difficult*

instance, figure 3.2a shows a curve that is nonlinear. The same data are plotted in figure 3.2b, where the y axis is logarithmic (that is, each increment is a multi-plication by ten—a linear scale is one in which each increment is an addition of one). Extrapolation of figure 3.2b is simpler than figure 3.2a and thus more likely to provide an accurate forecast.

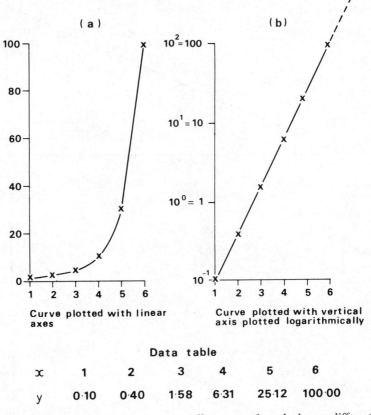

Figure 3.2 *One data set plotted on two different sets of axes looks very different*

Determination of the future demand is thus accomplished using existing data for the preceding years and is an attempt to ensure that needs for the next 30 years, say, will be met. If the yield required from the reservoir is greater than the inflow into it in a given time period then the water level in the reservoir will drop; it will rise again when inflow exceeds the yield taken. The reservoir must be designed to withstand low flows that can be reasonably expected to occur. Very low flows occur rarely. The flow that occurs, on average, once over a long but finite time span must be determined. This time period (the *return period*) is usually taken to be 100 years, which is considered to provide a sufficient safety margin. The size of the reservoir will thus depend upon the yield needed and the expected values for the smallest inflows.

One much-used technique for determining the size of the reservoir uses a *3-year synthetic* (or *design*) *drought*. This method takes all the flow data available for the reservoir inflow and, by extrapolation techniques, calculates the low flow that is likely to have a return period, T_r, of (usually) 100 years; in other words, this low flow value is likely to be observed, on average, once every

100 years. Although the probability that the flow actually occurs in a given year, P, is 1/100 it is not impossible (though it is unlikely) for the same low flow to occur in two consecutive years. (The same argument applies to high flows, or *floods*, of a given return period.)

Synthetic or Design Drought

To construct a synthetic drought for a period of 3 years it is first necessary to break down the 3-year period into 6 periods of 6 consecutive months: 3 summers and 3 winters. Summer is defined as the driest 6 months derived using data averaged over at least 30 years. This season is usually found to be May to October (the second half of the official 'water year'), but can sometimes be April to September. For each year of the record, the total 6-month summer flow is calcu-lated. If these are arranged in ascending order of value—known as *rank order*, denoted by r—then a return period is given by $T_r = (n + 1)/r$, where n is the number of flow records. The flows (or *discharges*) may be plotted against the return periods or against their probability of occurrence (= $1/T_r$), resulting in a curve similar to that in figure 3.3. In this case, re-plotting data on a logarithmic

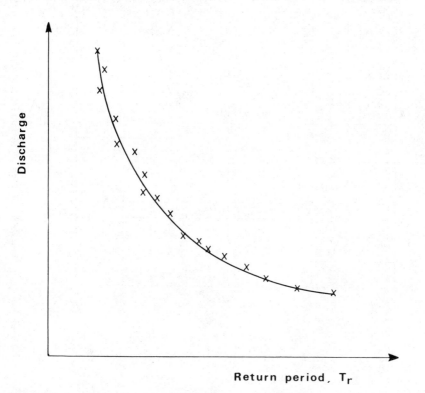

Figure 3.3 *Low flows plotted against their return period; extreme events occur only rarely (long return periods)*

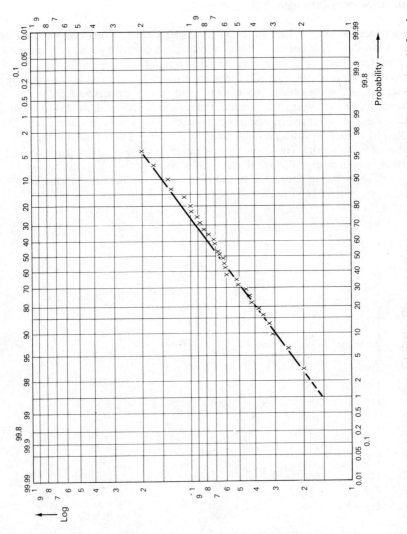

Figure 3.4 *Low flow data for 30 years plotted on log-probability paper; the straight-line fit to the data is also shown*

scale (cf. figure 3.2) does not help. However, it is found that if the values are plotted on specially drawn 'log–probability' paper, then a close approximation to a straight line results (see figure 3.4). This paper has a logarithmic vertical axis and a horizontal scale derived from a probability distribution known as the normal or Gaussian distribution (see the suggested reading at the end of this chapter for specific works on the subject).

The data, which might cover a range of return periods from 2 to 30 years, corresponding to a range in probabilities from 0.5 to 0.03 (that is, percentage probabilities from 50 to 3 per cent), can be extrapolated by a straight line to a probability value of 0.01 (1 per cent); see figure 3.4. This is the lowest flow (for a 6-month summer period) that is likely to occur once in 100 years.

To complete the drought, values for the low flow for a 12-month period and then 18-, 24-, 30- and 36-month periods are found (see figure 3.5). Subtracting the 6-month flow (A) from the 12-month flow (B) gives the flow likely during the first winter of the drought. Successive subtraction gives 6 6-month periods of low flow, labelled $S_1, S_2, \ldots, S_6$ in figure 3.5. As expected, successive summers

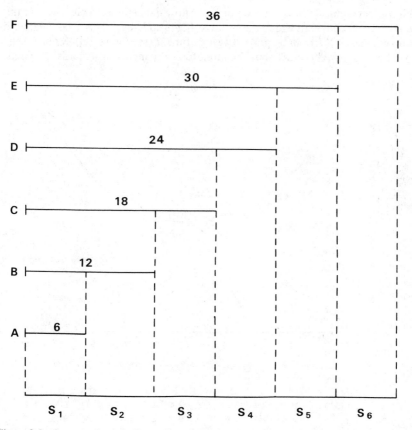

Figure 3.5 *Construction of a 3-year drought period, broken down into 6 6-month periods:* S_1, S_2, S_3, S_4, S_5 *and* S_6

become wetter, as do successive winters, that is, $S_5 > S_3 > S_1$ and $S_6 > S_4 > S_2$.

From the data, the ratio of the mean monthly flow for each calendar month to the mean value for its corresponding season can be calculated. These are denoted by R_i (i = 1 to 12). Multiplying the first summer value S_1 by $R_5, R_6, \ldots, R_{10}$ in turn gives values for the monthly flows in that first summer for the months May, June, . . ., October, respectively. The monthly flow for each month of the 3-year synthetic drought can thus be determined. These values are now used as the input to the reservoir model so that the effect on the water level in the reservoir throughout this severe drought can be studied. If the reservoir still contains water when the level is lowest then the yield can still be satisfied under these extreme conditions. The maximum depletion thus gives a good indication of the minimum size of reservoir that must be built. Two methods are in current use—a graphical technique using a *mass curve* and computer simulation.

Mass-curve Technique

This uses a graphical representation of the inflow data (for example, the monthly values derived above) in which the inflows, I_i, are added up. This accumulated inflow (written as ΣI_i) can be plotted against time, t (see figure 3.6). Since I_i can never be negative, this curve must be monotonically increasing (that is, ΣI_i must

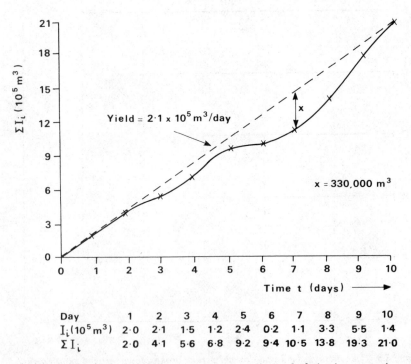

Day	1	2	3	4	5	6	7	8	9	10
$I_i (10^5 m^3)$	2·0	2·1	1·5	1·2	2·4	0·2	1·1	3·3	5·5	1·4
ΣI_i	2·0	4·1	5·6	6·8	9·2	9·4	10·5	13·8	19·3	21·0

Figure 3.6 *Mass-curve technique to determine maximum depletion in a reservoir*

never decrease). If the total amount $\Sigma\, Y_i$ taken *out* of the reservoir by a time t in order to satisfy a yield $\bar{Y}$ (= $\Sigma\, Y_i/T$ where T is the length of the drought, i.e. 36 months) is then subtracted from this curve, then the depletion in the reservoir will be $\Sigma\, Y_i - \Sigma\, I_i$. The maximum value of this difference (labelled x in figure 3.6) gives the size of the reservoir needed. Some adjustment is necessary if any water is lost by spillage when the reservoir is too full; this is not likely to occur during a drought. As an additional safety factor, it is insisted that the reservoir shall be full both at the beginning and at the end of the drought. Then the sum of the yield over the 3-year period will equal the sum of the inflows, I_i (see figure 3.7). Thus, the total yield over 3 years equals the total inflow. Hence the maximum safe yield over the drought period is given by $Y_{\mathrm{safe}} = \Sigma\, I_i/T$. If the demand is greater than this, then the reservoir may fail.

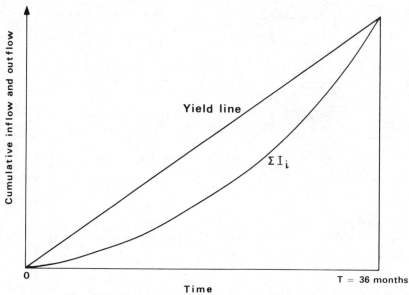

Figure 3.7 *Cumulative mass-curve typical of a 3-year drought*

This calculation has assumed that the reservoir is to be used in one specific way: to deliver a yield of potable water directly to the consumer. This is called the *direct supply mode* of operation. The yield over a given period is simply equal to the amount entering the reservoir during that time. However all *impounding* reservoirs and even many of the *pumped storage* reservoirs are built on an existing stream so that the minimum acceptable flow must always be maintained in the stream, thus decreasing the amount of water available for consumption. Since the stream is fed solely by the reservoir, then the M.A.F. is a constant value and so the total yield (that is, the reservoir output) equals the sum of the M.A.F. and the drinkable supply. Instead of using a graphical technique it is a simple procedure to employ a computer technique. The logic of both the graphical and numerical techniques may best be illustrated by means of a *flow chart* (see figure 3.8). On this flow chart, the possibility is included that the reservoir overflows at some time step and water is lost.

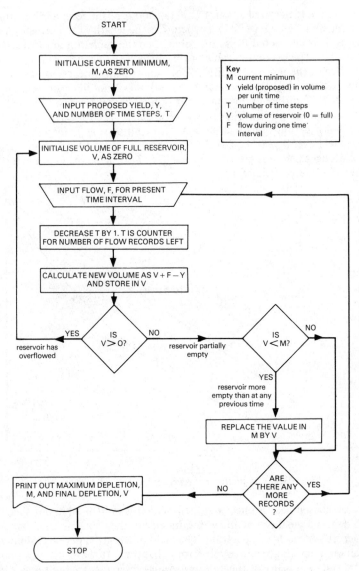

Figure 3.8 *Flow chart for direct supply reservoir*

The alternative method of reservoir utilisation (the *regulating mode*) can be modelled by an iterative procedure, also well suited to computer solution. A regulating reservoir is built on a tributary. At times of low flow in the main stream, water is released from behind the dam, and soon flows down the tributary to enter the main river and enhance its flow. Thus the natural flow (which may well be less than the M.A.F.) is augmented, so that the yield can be abstracted from the river without impairing the stream's natural characteristics. Figure 3.9 illustrates this.

CATCHMENT FOR RESERVOIR

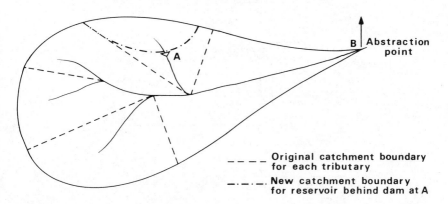

Figure 3.9 *Location of a regulating reservoir at point A; it has a smaller catchment than a direct supply reservoir at point B because of its position on one of the tributaries*

The dam is at point A, whereas the water is abstracted from the river at point B. The M.A.F. must be maintained at this point of abstraction. Care must be taken to allow for the time lag of water released from A to travel down the river bed to point B.

The reservoir at A is now only collecting direct runoff from a small part of the catchment (possibly only 10 per cent) compared to the 100 per cent collection of an impounding reservoir at the point of abstraction (B). Since the amount of water now needed to maintain riparian rights may alter between zero and the whole of the M.A.F. it is no longer simple to deduce what fraction of the total yield of the reservoir is for river regulation and what proportion for consumptive supply; the maximum (potable) yield must be guessed. Allowing for river regulation, the 3-year drought scheme can be implemented, as before. If the reservoir is full at the beginning and end of the 3 years, then the guessed yield is correct. If not, as is more usually the case, then another guess must be made. If the reservoir was overfull at the end, then the yield must be more and vice versa. The flow chart to assist in the solution of this problem is shown in figure 3.10.

To compare these two types of usage, table 3.1 shows the yield, M.A.F., supply and volume for a reservoir at point A: the first line illustrates its use in the regulating mode and the second line its use in the direct supply mode. (The numerical values quoted are ratios and therefore have no units associated with them.) Thus

Table 3.1 Comparison of Regulating and Direct-supply Modes for a Reservoir

Mode	Yield	M.A.F.	Supply	Volume of reservoir
Regulating	15	11 (at B)	4	60
Direct-supply	3	1 (at A)	2	32

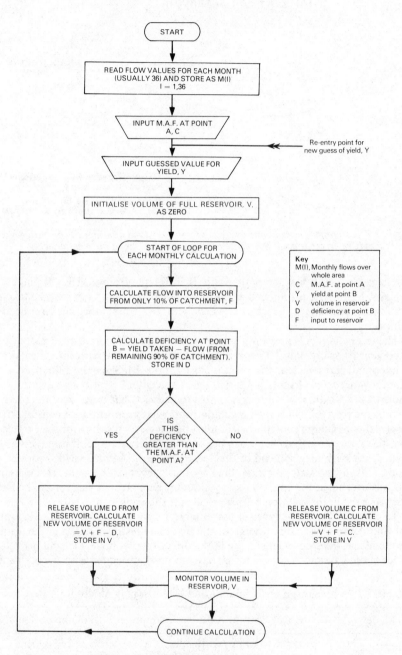

Figure 3.10 *Flow chart for a regulating reservoir*

doubling the reservoir size will guarantee twice the potable supply *only* if the reservoir is used in a regulating mode.

However, in this case, since the water is abstracted well downstream it contains more pollutants and hence the cost of treatment is higher. Some benefit can be gained from this by permitting the use of the lake for recreational purposes; maintenance of purity is then no longer an overriding priority.

The size and yield of the new reservoir, which depend upon the inflows during the artificial drought, are thus calculated. This gives some indication to the engineers of the size of valley needed. Geological surveys must be undertaken to determine the viability of any proposals and an economic evaluation of the chosen location must be made, including engineering costs, transportation costs, rehousing and compensation to the displaced inhabitants.

Summary

The determination of the size of reservoir needed to meet increasing demands on the water supply of an area is based on the forecasting of future events. Both the forecast demand for the area and the predicted extreme low flow event (drought) must be derived. Future demands are predicted by extrapolating curves plotted for the past years, and some attempt is made to take into account fluctuations in population and standard of living.

Drought prediction is again based on past records. A frequency of occurrence of once in 100 years is usually considered rare enough for design purposes. Since most flow records are considerably shorter than this, a synthetic 3-year drought based on nonlinear extrapolation techniques is often used. This synthetic drought can then be used as inflow data for the reservoir and the capacity of the reservoir that is necessary in order to give the required (reliable) yield over the whole 3-year period can be calculated. The size is found to depend upon the mode of operation of the reservoir (regulating or direct supply), which may be determined, in part, by economic considerations.

Suggested Reading

Arthurs, A. M., *Probability Theory* (Routledge & Kegan Paul, London, 1965).
Twort, A. C., Hoather, R. C. and Law, F. M., *Water Supply*, 2nd ed. (Edward Arnold, London, 1974).

4

Dam Design

The size of reservoir needed so that postulated future demands are satisfied depends on the magnitude of this demand, the hydrological features and the funds available for the project. Other factors that will influence the choice of reservoir site include proximity to demand centres, quality of the stored water and the biota that will thrive in the lake. These points are all discussed in detail in other chapters.

Although the costs involved in the project (see chapter 5) will be largely determined by the scale of the work, calculations must also take into account the various technologies that will be used by the civil engineering contractors (for example, earth or concrete dam). This choice will be dependent on the geology of the chosen site. The ground must be capable of supporting a large mass of water, without excessively large losses by seepage. It must be technologically possible to build a dam across the valley and the feasibility (and cost) of all necessary demolitions, road rerouting, etc., must be considered.

Geological Survey

The first consideration is the size of the reservoir that is needed. A short list of suitable valleys in the specified area is a starting point for more detailed topographical and geological considerations. Ordnance Survey 1:25 000 maps quickly pinpoint possible dam sites—for example, where a constriction occurs in the valley. Maps produced by H.M. Geological Survey indicate whether there are any major faults in the valley and whether the base is likely to be watertight. At this stage too, the number of dwellings in the area can be established, so that costs of rehabilitation may be included in the calculation. From these map investigations several locations will be chosen for site surveys; the investigations must cover not only the lake area, but also any associated diversion works, the gathering grounds (for an impounding reservoir) and, perhaps most importantly, the dam site itself. The geology of the area determines whether the lake basin will be naturally watertight—an alternative is concrete lining. *Cut-off works* to prevent leakage are often necessary, especially at the downstream end. A *grout curtain* (cement filling) may be introduced under the dam to prevent seepage (see later discussion). It is not only the surface features that must be investigated. The type of rock tens and even hundreds of metres below the surface, together with knowledge of its permeability, must also be considered. Water pressure in rocks and soils can be measured by sophisticated piezometric apparatus. In spite of the prevalent concern for watertightness, groundwater levels have been severely neglected in the past. When the reservoir is full, changes in the water table may be sufficiently large to cause concern to local farmers and householders, and stream levels may be altered.

Investigations of the bedrock can be carried out in several ways: measurements of electrical resistivity can be interpreted in terms of rock porosities, seismic methods give information on many aspects of the lithology, while magnetometers provide a means of identifying the basic geological structure. Perhaps the most useful method is *borehole logging*, in which a hole is drilled vertically in the rock and either the core is inspected or else photographic methods are employed within the vertical shaft. Soil samples, usually from a core of 102 mm (4 in.) diameter are removed and analysed.

Even with careful examination of the area, mishaps do occur. The final site chosen for the new storage reservoir in the East Midlands was a valley near the village of Empingham. The reservoir there was filled in 1975-6. During the summer of 1977 it was found that large volumes of water were leaking out of the lake (Rutland Water) through the bedrock. Initial attempts to trace the leak failed and almost a year later no solution had been found.

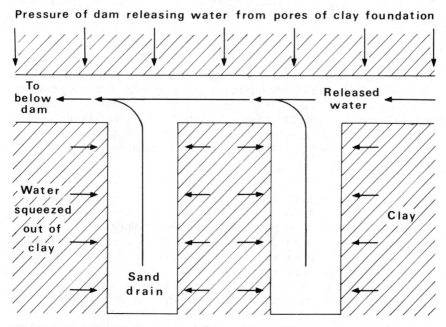

Figure 4.1 *Sand drains inserted in clay base beneath dam to remove excess water 'squeezed' out of the ground*

The valley slopes above and below the dam are also examined to discover whether they have been subject to any major disturbances in the past. The safety of an old slip that has stabilised itself may well be endangered when excavations begin. Recently, work was initiated to increase the water supply to Bradford by increasing the height of the concrete dam of Grimwith Reservoir. When this reservoir had first been constructed (in the mid-nineteenth century, by the same engineer as the Dale Dyke Dam), there had been problems. Geological faults were known to exist and the initial construction had often been criticised. Nothwithstanding, the reservoir had many decades of faultless service to its credit and thus the lesson of history was ignored. During reconstruction work in 1976, a major slip occurred in the valley walls below the dam. Steel pins, several metres long, had to be driven into the valley sides to secure the strata, thus delaying work by several months.

Water can again be a problem when the dam itself is added, especially if the ground is clay. Geological surveys will reveal how much water is held in the pores of the soil. If saturation levels are approached, additional pressure (from the weight of the dam) may squeeze the water out of the pores on to the surface but below the superposed dam—in that case, the dam would float away! To obviate this, many such dams have *sand drains* built into the foundations (see figure 4.1). These collect the excess water and channel it away to safety.

Types of Dam

Based on the findings of the geological survey, the dam site will be chosen. Two basic types of dam exist

(1) earth dams
(2) concrete (including masonry) dams.

A concrete dam requires stronger foundations than an earth dam since the latter tends to be broader and its weight is thus spread more evenly over a larger surface area. If the foundations settle, then an earth dam will accommodate this disturbance, whereas a concrete dam tends to fracture. Choice of the type of dam is seldom found to be arbitrary; there is often a strong preference for one type, dictated by the geological survey or the contractor's expertise.

Earth Dams

In building an earth dam, the design of the cross-sectional shape of the dam will have to take into consideration the properties of the foundation material, the fill material and also the time available and the weather conditions. This latter is especially important if clay is to be used, as it very often is in Britain. In this basic design concept, the central core will be of rolled clay (replacing the puddle clay previously used), although a small number of earth dams contain a concrete core. On either side of the core there will be embankments of clay fill, with additional semipervious fill in the outer regions. The upstream face is usually protected by *riprap* (dumped quarry rock, often limestone), so as to reduce erosion by waves.

The stability of the dam used to be a result of careful and slow workmanship, augmented by experience. Today, developments in soil science allow preliminary numerical calculations to be undertaken at the design stage. If the predictions are satisfactory (once any recommended safety factors have been incorporated into the design), work on the dam can proceed with safety and speed. Slip-circle analysis is often used, in this technique, various slip circles are analysed to find the critical case. (It is usual to include an additional safety factor of 1.5.)

Rock fill at the toe of the dam may assist stability. Watertightness is achieved either by a *cut-off trench* (which is a downwards extension of the central water-tight core) or by *grouting*—a technique that is regularly used to reduce the permeability of rocks: the addition of cement, clay or chemicals to ensure that the rocks present an impenetrable barrier. Grouting was first used in 1876 (at the Turnstall Reservoir, Durham), but this was only to stop a leak, and the first application of a grout curtain at the initial stages of construction did not occur until the 1920s—at the Scout Dike Reservoir and the Bartley Reservoir. In current practice, the grouting material is introduced into holes 44 mm in diameter drilled into the rock strata. Liquid cement is then pumped in under pressure and solidifies in the cracks and fissures in the rock. The necessity for grouting varies; from cases in which it is completely unnecessary through Claerwen (requiring 6.7×10^4 kg or 66 tons) to Ladybower, which needed 1.6×10^7 kg (16 000 tons) of grout material.

Concrete Dams

There are three basic types of concrete dam: gravity, arch and buttress (see figures 4.2, 4.4 and 4.5, respectively). The choice between these depends upon the cost and feasibility. Estimates for the cost of the Glen Shira Dam for rockfill,

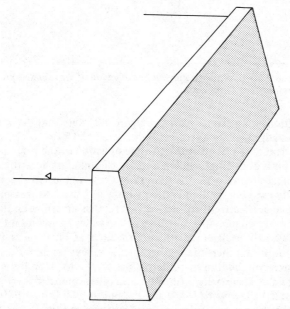

Figure 4.2 *Gravity dam*

gravity and buttress dams were in the ratio 1.4:1.25:1. An arch dam could not be considered since it requires a valley where the ratio of the dam length to height is significantly less than 5. (Few such valleys exist in Great Britain.) The advantages of a concrete structure over an earth dam are that the design is more precisely determined, the construction materials are more stable and outlet pipes can be built into the structure safely. Flood waters can pass over the dam without fear of erosion.

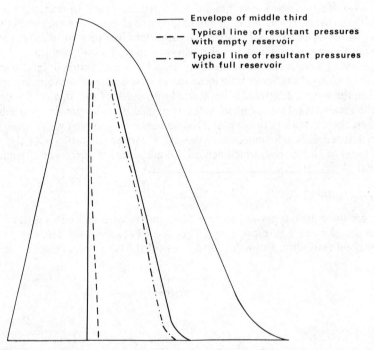

—————— Envelope of middle third

— — — Typical line of resultant pressures with empty reservoir

—.— Typical line of resultant pressures with full reservoir

Figure 4.3 *For stability of a gravity dam the line of action of the resultant pressure must lie within the middle third of the dam*

Safety conditions for a *gravity dam* (see figure 4.2) state that the maximum compressive strength must not exceed safe limits (64 to 150 x 10^4 N/m^2). To ensure that the dam does not overturn, the line of action of the resultant pressures when the reservoir is both empty and in a state of flood must lie within the middle third of the dam (see figure 4.3). The third safety criterion insists that it should not slide, possibly carrying the foundations with it. For aesthetic reasons, some gravity dams (such as Claerwen) are curved, but this has no effect on the stability.

In an *arch dam* (see figure 4.4), the force of the water is largest on the side walls of the valley, as a result of the effect of the arch, and may be as high as 7×10^6 N/m^2. In Britain there are few acceptable valleys for arch-dam construction. The first arch dam built in this country was across the Afon Prysor, one of four restraining Lake Trawsfynnyd (in Wales). In other countries there are larger numbers. The Bear Valley Dam (1884) and the Buffalo Bill Dam (1910) in the United States, the Parramatta Dam (1852) in Australia and the Kariba Dam in

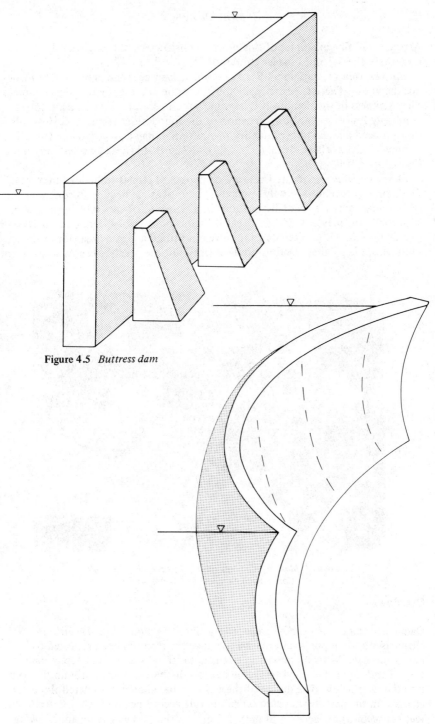

Figure 4.5 *Buttress dam*

Figure 4.4 *Double-curvature arch dam*

Africa are all fine examples of this type of construction, although Kariba does not satisfy the 5:1 ratio for length:depth.

Buttress dams (see figure 4.5) have several advantages and are common throughout the world. There is virtually no uplift pressure and they require less concrete. The thickness of the dam itself is less, and so temperature variations are correspondingly lower. Access for inspection of the foundations is possible. However, they do need a much larger quantity of shuttering than a gravity dam. Typical examples exist at Haweswater (1934–41), at Lamaload (near Macclesfield) and at Clywedog in Wales (1966).

Masonry dams also exist. These are composed of stones embedded in mortar. The term has also been loosely applied to dams of stone bedded in concrete, and so masonry dams are generally included in discussions of concrete dams. The most important true masonry dam in Great Britain is the Vyrnwy Dam built in 1881–91 (see figure 4.6). Other such dams followed in Britain and America (for example, Thirlmere, Elan Valley, Derbyshire Derwent), but the practice is now obsolescent.

Figure 4.6 *Masonry dam restraining Lake Vyrnwy in Wales (Courtesy of Severn-Trent Water Authority)*

Overflows

Dam construction is usually preceded by river-diversion work. Diversion channels are often put to alternative use once the river has been returned to its natural course. Provision must also be made for flood water. In concrete dams a weir or *spillway* is provided to channel excess discharges over or around the dam into the river below. The design of these is now based on extrapolated flood discharges. In the past it was based on an old rule according to which $\frac{1}{2}$ cusec (cubic feet per second) per acre (in SI units 3.5 m^3 s^{-1} km^{-2}) was assumed, allowing for 0.9 m (3 ft) of head over the weir. This led to a weir with a length of

Figure 4.7 *Bellmouth overflow (Rutland Water).*

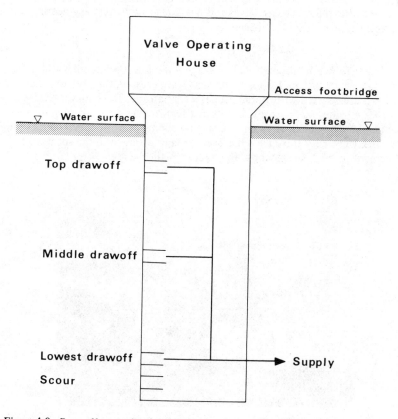

Figure 4.8 *Drawoff tower for the abstraction of water for a direct supply reservoir*

0.9 m per 4.05×10^5 m² (40.5 hectares or 100 acres) of gathering ground.
An alternative method for flood-water disposal is a *bellmouth overflow*
(see figure 4.7), built within the limits of the reservoir itself; this structure
channels overflows away. Under normal conditions the bellmouth is isolated from
the body of the water in the reservoir by a protective wall. Under flood conditions
both this wall and the bellmouth (which is slightly lower) are topped and the water
removed safely. Although in regular use—for example, at Rutland Water, Ladybower, Llyn Celyn in Wales and Balderhead—it must be remembered that a bellmouth has a finite maximum capacity. Thus accurate evaluation of maximum
likely floods is important at the design stage.

If water is to be abstracted directly from the reservoir, there must be a *draw-
off tower* (see figure 4.8) within the lake, and pumps and a pumping station will
also be necessary to transfer the water to the treatment works. In a pumped-
storage scheme additional tunnelling and pumping stations must be built to
collect and transport water that is to be input to the reservoir.

Erosion

Especially on larger reservoirs, where the *fetch* is substantial, wind action may be
an important factor in causing waves and possible erosion. One engineering formu-
la predicts the wave height in metres as

$$0.34 \sqrt{F} + (0.76 - 0.26 \sqrt[4]{F}) \tag{4.1}$$

where F is the fetch in kilometres. For a lake 9 km long ($\approx$ 5.5 miles) the wave
height may be as much as 1.3 m. To prevent wave erosion, the upstream faces of
earth dams are covered with riprap usually of limestone. The banks themselves
may suffer erosion. Bank erosion can even affect smaller-scale lakes where wind
action is minimal—ducks often enter a lake by sliding down the bank and eroding
it. At York University this factor has been a major problem to the successful
management of the lake system in the middle of the campus.

Figure 4.9 *Normanton Church (Rutland Water)*

Although not directly concerned in the construction of the reservoir itself, peripheral civil-engineering contracts may be necessary. These will often involve rerouting flooded roads and such diverse matters as filling the foundations of a church (for example, Normanton Church in Rutland Water—see figure 4.9) and removing ancient monuments (as at Nubia for the Aswan Dam). Water mains, electricity and telephone lines must also be rerouted.

Summary

Once the reservoir size has been determined, a suitable site must be found, usually within the vicinity of the demand centre. The geology of the area must be thoroughly investigated; any valleys that are geologically unsound (for example, they may have geological fault lines) would be too costly, since extra concrete would be needed to ensure the reservoir was watertight. The type of dam chosen will be either earth or concrete, a choice that may be dictated by other external factors. The design of the dam must ensure stability against fracture, overturn and sliding. It must also be able to withstand flood levels. The overflow in the reservoir should be capable of safely removing large volumes of water from the reservoir. Especially in reservoirs exposed to the wind, protection against erosion by wind-generated waves may be necessary, probably in the form of a stone blanket on the upstream face of the dam. Similarly the banks themselves may need protection against wave action. Other civil-engineering contracts may be necessary and may be of a very specialised nature (for example, the removal and relocation of churches and other ancient monuments).

Suggested Reading

Institution of Water Engineers and Scientists, *Manual of British Water Engineering Practice,* ed. W. O. Skeat, 4th ed. (Heffer, Cambridge, 1969).
Thomas, H. H., *The Engineering of Large Dams* (Wiley, New York, 1976).

5

Economics

To satisfy water demands, reservoir capacities can be calculated so that the lake
runs dry only once in n years (see chapter 3). Choice of reservoir site is restricted
not only to valleys capable of containing the required volume, but also to those
whose use will cause least upheaval in the lifestyles and livelihoods of the local
populace. The geological structure of the valley must be investigated. All these
factors bear upon the economic considerations of choosing the best site for the
new supply. These ideas should be applied not only to the initial consideration of
the capital costs of construction but also to the running costs.

Costs and Benefits

The *costs* of water supply schemes can be assessed using standard economic
theories. The *benefits* accrued from the scheme are much harder to assess quantit-
atively and are often ignored in planning future water supply schedules.

To compare the capital costs of alternative water supply schemes, *discounting*
is introduced. In a construction project that will last for several years, allowance
can be made for capital tied up in construction costs, which would otherwise be
available for investment. This is done by relating all costs to the *present value*
(P.V.) by using a *discount rate*. If the discount rate is r (= 100 × r per cent) then
an investment of £c would be worth £c $(1 + r)$ after one year. After i years, its
value would have risen to £c $(1 + r)^i$. Reversing the procedure, it can be seen that
a capital cost of £c $(1 + r)^i$ in the ith year would be available from an initial
investment of £c. This value, £c, is the required P.V. for the expenditure in the
ith year; this is often expressed—dividing by $(1 + r)^i$—as follows: 'The P.V. of a
capital cost of £c_i in the ith year is £$c_i/(1 + r)^i$.' In recent years, the standard
value for the discount rate, r, used by the Water Authorities has been 10 per cent.

For projects that take several years of construction plus operation in order to
reach maximum operational capability, a *discounted unit cost* (D.U.C.) can be
calculated as the uniform price chargeable each year such that the P.V. of the
income would be balanced by the P.V. of the outgoing costs. For a discount rate
of r and a project life of n years, this unit cost, p, can be expressed as

$$p = \frac{\sum\limits_{i=1}^{n} c_i/(1+r)^i}{\sum\limits_{i=1}^{n} q_i/(1+r)^i} \qquad (5.1)$$

where c_i and q_i are the total costs and the yield of the project in the ith year. This calculated unit cost is found to be useful in comparisons of alternative schemes.

There are also several less quantifiable aspects of such a project that must be taken into account. These are the costs and benefits to the public (especially those resident in the locality) that may occur either during the construction phase or when the reservoir is in full operation. The cost of rehousing people living within the lake boundary is easily taken into account. Less easy to assess is the decrease in the visual 'environmental value' caused, either on a short time scale by contractors' vehicles, or on a longer time scale by the reservoir and its associated treatment works. (These are often known as *social costs and benefits* or as *shadow costs and benefits*.)

Visual amenity is strongly subjective, although recently attempts to quantify the visual qualities of locations have been presented in the literature. The intrusion of a reservoir into a landscape may become a benefit to the area both in terms of increased tourist attraction and in the possible increased inflow of cash from visitors to local tradesmen.

The costs and benefits to the consumers (who are not necessarily the people living in the vicinity of the new reservoir) must also be assessed.

Running Costs

The cost to the consumer would, ideally, defray the expense of maintaining the new source. This cost can be assessed in two ways: on either a long-term or a short-term basis. For a reservoir that can supply the community at a maximum

Table 5.1

Output, Q		Total fixed cost, T.F.C.	Total variable cost, T.V.C.	Total cost, T.C.	Average total cost, A.T.C.		Marginal cost, M.C.
(m³/s)	(m.g.d.)				(£/cumec)	(£/m.g.d.)	
0.5	9.51	1000	20	1020	2040	107	20
1.0	19.01	1000	40	1040	1040	55	20
1.5	28.52	1000	60	1060	707	37	20
2.0	38.02	1000	80	1080	540	28	20
2.5	47.53	1000	100	1100	440	23	20
3.0	57.03	1000	120	1120	373	20	20
3.5	66.54	1000	160	1160	331	17	40
4.0	76.05	1000	240	1240	310	16	80
4.5	85.55	1000	400	1400	311	16	160
5.0	95.06	1000	720	1720	344	18	320

rate of, say, 5.0 m³/s (95.06 million gallons/day (m.g.d.)), calculation of the *short-run cost function* must include the running costs, which consist of two components: *total fixed costs* (T.F.C.) (such as labour, regular maintenance), which are necessary whether the reservoir is running at full operation (5.0 m³/s) or only at 0.5 m³/s, say, and the *total variable costs* (T.V.C.), which would include pumping costs, directly related to the output. The *average total cost* (A.T.C.) per unit of water is thus the sum of T.F.C. and T.V.C. divided by the output Q. Table 5.1 gives a numerical example in which the difference in total costs for one extra unit of water, the *marginal cost* (M.C.), is also shown. The marginal cost is

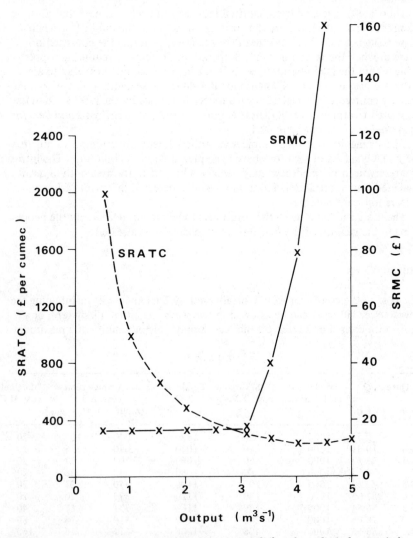

Figure 5.1 *Short run costs (marginal and average total) for a hypothetical reservoir for which the output data are given in table 5.1*

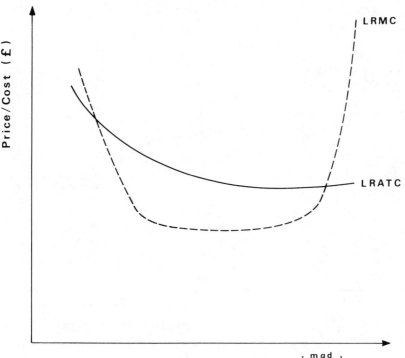

Figure 5.2 *Typical long run costs for a reservoir*

an important concept used to assess the extra cost of supplying more water. This is apparent in figure 5.1, where it can be seen that, as output increases, the *short run average total costs* (S.R.A.T.C.) fall to a minimum but begin to increase for desired outputs greater than 4.5 m^3/s (85.55 m.g.d.). At the same time the *short run marginal costs* (S.R.M.C.) also escalate. This illustrates that, beyond a certain point, supplying a large output tends to become 'uneconomic'.

A similar phenomenon occurs when the long-run costing is considered. However, in this calculation all the factors must be considered as variables, although present-day technology is assumed out of necessity. Figure 5.2 depicts typical curves for the *long-run average total costs* (L.R.A.T.C.) and the *long run marginal costs* (L.R.M.C.). Increasing output past a certain point again results in a rapid increase in marginal costs. This approach is useful in discussing the pricing structure that ought to be used. For instance, if the relationship between metered and unmetered supplies is investigated, then recommendation can be made as to the effectiveness both in minimising costs and limiting wasteful use. Figure 5.3 shows the total demand (domestic plus industrial) in a region as a function of price. (It is assumed that the domestic supply is unmetered[1] and all the industrial demand

[1] In Great Britain, only Malvern has a metered domestic supply, although the Fylde area was partially metered on an experimental basis during 1967–72. Later studies have been undertaken in Mansfield and in Peterborough.

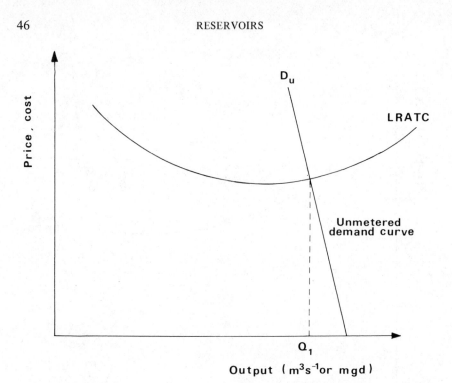

Figure 5.3 *Unmetered demand curve and long run average total costs define a demand Q_1*

is metered.) This curve (labelled D_u) reflects the fact that if the price of water falls then the demand will tend to increase slightly. However the cost of *supplying* that water has already been defined in terms of the L.R.A.T.C. curve. When these two curves intersect, total costs are balanced by total price paid by the consumer, which defines the demand as Q_1.

Metering

If the effectiveness of metering is considered, then the cost–benefit analysis is performed in terms of the long run marginal costs. The demand from consumers who pay for their water on a metered basis is usually found to be reduced. In Malvern a fall in demand of 10 to 20 per cent followed the introduction of metering (although more recent surveys—see chapter 1—cast doubts on these figures) and in Boulder City, Colorado a fall of 30 to 40 per cent was observed in domestic use. This can be illustrated using figures 5.4 and 5.5. In figure 5.4 the postulated demand curve for a metered domestic supply (curve D_m) indicates that, as prices fall, demands increase. The intersection of this curve with the L.R.M.C. curve defines a demand Q_2. To determine the saving introduced by metering, figures 5.3 and 5.4 must be replotted on the same axes—see figure 5.5. The marginal cost incurred by increasing the demand from Q_2 to Q_1 is simply the area under the curve between these points—that is, the area ABQ_1Q_2. This area thus represents

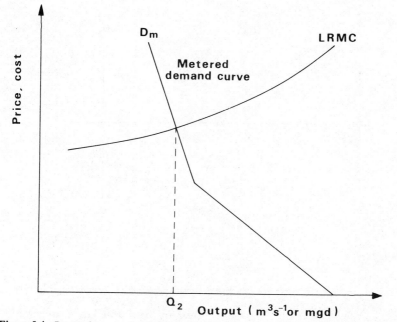

Figure 5.4 *Demand curve after the introduction of metering; the intersection with the L.R.M.C. curve defines the demand Q_2*

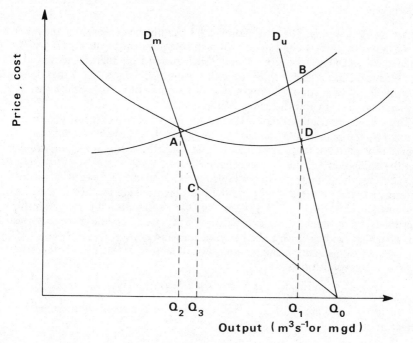

Figure 5.5 *Demand curves and demands before and after metering*

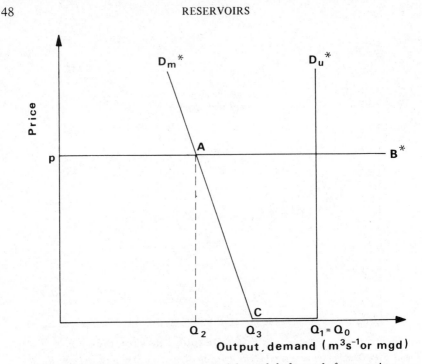

Figure 5.6 *Simplified demand curves and demands before and after metering*

the saving by metering. The cost with which this must be compared is the net loss of consumer surplus (which can be shown to be equivalent to the area $Q_2 ACQ_0$ minus the area $DQ_1 Q_0$) plus the direct costs incurred from meter reading and maintenance. Recent calculations using this technique seem to show that domestic metering is now an economic proposition in a number of areas in South-east England and in many new housing developments.

This calculation can be simplified since the marginal costs are not significantly affected by the reduction in demand (from Q_1 to Q_2). Curve B can then be replaced by a horizontal straight line, B* (see figure 5.6). If it is further assumed that the fall in demand due to metering itself occurs instantaneously (from Q_1 to Q_3) and that the unmetered supply is also inelastic (curve D_u*) then the saving by metering is given by $p(Q_1 - Q_2)$ and the costs incurred are the sum of the loss of consumer surplus, $\frac{1}{2}p(Q_3 - Q_2)$ (assuming D_m* represents a linear relationship); consumers' inconvenience and expenditure on repairs to avoid unnecessary wastes, etc., E, which can be shown to be between zero and $p(Q_1 - Q_2) - \frac{1}{2}p(Q_3 - Q_2)$; and the cost of installation and maintenance of meters, M. Metering is therefore an economic proposition if

$$p(Q_1 - Q_2) \geqslant M + E + \tfrac{1}{2}p(Q_3 - Q_2) \tag{5.2}$$

If consumers' costs are ignored, as they may well be as a result of inadequate data, this criterion becomes

$$p(Q_1 - Q_2) \geqslant M \tag{5.3}$$

Consumer Surplus and Marginal Utility

The *utility* of a piped water supply is related to the amount of 'satisfaction' derived by the consumer. The *total utility* measures the total satisfaction from *all* units that are used in a given period, whereas the *marginal utility* is the extra satisfaction derived by increasing consumption by one extra unit (cf. marginal costs discussed above). If a demand increases and can be satisfied, then the price per unit volume of water tends to fall (see table 5.2). If the price is plotted against

Table 5.2

Price per 1000 m³ (pence)	5	4.5	4	3.5	3
Demand per week (1000 m³)	5	6	7	8	9

the demand, then the extra satisfaction derived by increasing demand is represented by the area under the stepped curve (see figure 5.7). The sum of the marginal utilities gained by increasing demand by 5 units is thus the sum of the 5 rectangles. Further increasing demand adds less and less to the satisfaction derived. Thus the necessity for the Water Authority to acceed to satisfying such high demands could reasonably be questioned.

The curve of price against demand is often known as the *demand schedule*. In figure 5.8, the present consumption is assumed to be given by Q_A and the unit price is P_A. The total price paid is thus $P_A \times Q_A$, which is the area $OQ_A AP_A$ in

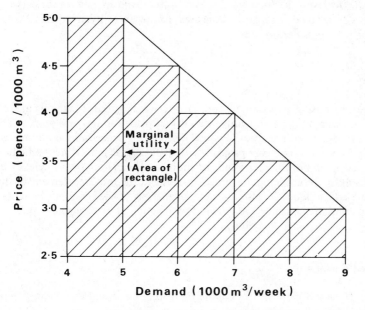

Figure 5.7 *Demand schedule for data in table 5.2*

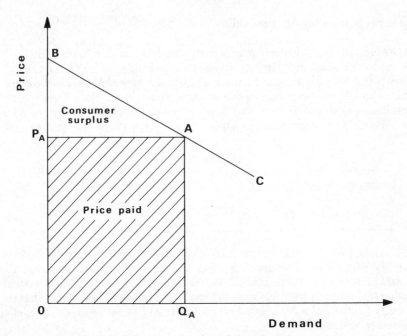

Figure 5.8 *Demand schedule and consumer surplus; the total worth of the water is given by the area under the curve (OBAQ$_A$)*

figure 5.8. However, the total worth of the water is the area under the curve (area OBAQ$_A$). The difference in these areas (represented by area BAP$_A$) is known as the *consumer surplus*.

Summary

The benefits of using an economic approach to water resource planning are several. Decision-making on the comparative viability of plans for new reservoirs can be assisted by cost–benefit analysis in which discounting procedures are found to be necessary in order to compare projects with differing cash investment schedules. Running costs (in both the long and the short term) can be shown to yield a useful criterion for determining the level of prices and the resulting demand at that price. These techniques are also useful in an analysis of the economic benefits of metering domestic consumers and seem to indicate that metering is, at present, a viable alternative in many regions.

Suggested Reading

Bierman, H., and Smidt, S., *The Capital Budgeting Decision* (Macmillan, New York, 1960).

6

Management of Reservoirs

Finding the most economic way to distribute water from a set of reservoirs in order to satisfy the varying demands within a distribution area introduces the problem of *optimisation*. The reservoir manager is able to vary the withdrawal rates from the several reservoirs under his control so as to take into account the location of demand, pumping costs and the level of water in each reservoir. Automation, both in data collection and in control, is being introduced to help in running many of the newer and/or larger supply networks.

Supply Networks

The network of feed lines within a given location may be complex (see figure 6.1). Optimisation at every stage may not be possible, since the temporal variation of demand patterns imposes severe constraints on the problem. The two main requirements, however, are to satisfy the demands of the consumer at one end of the supply network, and to ensure that the sources of supply do not fail.

The information available to the controller of such an operation is unlikely to be based solely on the collection of data. Especially at the consumers' end, demands are likely to be *forecasts*: extrapolations of present data and past trends. Where more detailed data are available, these are more likely to be retrospective (from surveys) and useful only for long-term strategic planning. Data on daily variations in both supply and demand are less easily obtained.

Knowledge of the changes in the volume of water available for storage is usually gained by direct measurement. In the Empingham scheme, water is abstracted from the Rivers Welland and Nene (at Tinwell and Wansford, respectively) when these are in flood. Metering at the abstraction point is automatic and the data are transferred by a *telemetry* link to the control station at Empingham. Similar expensive and technologically advanced communication systems have been installed at Falmer near Brighton. Telemetry systems are also used by the Lower Ouse Division of the Anglian Water Authority and in the control of Amsterdam's water supply, resulting in greater sensitivity of control for the Water Manager.

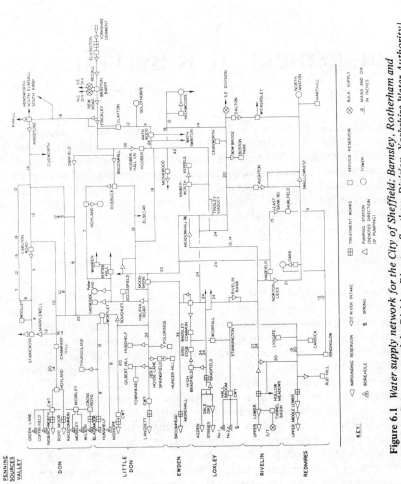

Figure 6.1 *Water supply network for the City of Sheffield; Barnsley, Rotherham and Rawmarsh (courtesy of the Division Director, Southern Division, Yorkshire Water Authority)*

Distribution Problems.

However complex the distribution system is, a first attempt at optimisation can be made by reducing the network to a set of suppliers (reservoirs) and a set of consumers (townships), which can be interlinked (reservoir to township) in any number of ways. A simple example of this is illustrated in figure 6.2, in which the reservoirs supply three demand centres. Four pipelines are used, along which there is a flow of x_i (m^3/day) at a cost of c_i (in units of, for example, £10 000/m^3/day). Using the techniques of *operational research*, it is possible to deduce the cheapest way of ensuring that the demands are satisfied and the reservoirs do not fail (that is, become empty).

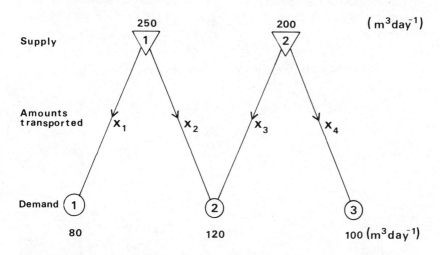

Figure 6.2 *Typical interlinking between several supplies (such as storage reservoirs) and several demand centres (such as townships)*

To find this solution it must be assumed that the reservoirs already exist (that is, there are no capital costs involved); that pumping costs are proportional to the flow rate; that forecast demands are accurate; and that transmission is by pipeline (that is, there are no evaporative losses). The cost of local distribution must also be neglected. In this example, the demands of the three towns are 80, 120 and 100 units respectively (1 unit = 1000 m^3/day) and the two reservoirs are able to supply a maximum of 250 and 200 units of water, respectively. In principle, there is sufficient water available to satisfy all demands; but should township 2 receive all its water from reservoir 1 or from reservoir 2—or from both? The total cost of pumping is

$$\sum_{i=1}^{4} c_i x_i \ (= c_1 x_1 + c_2 x_2 + c_3 x_3 + c_4 x_4)$$

and must be minimised within certain *constraints*. The constraints of the problem are simply our two basic assumptions: demands must be satisfied (criterion A) and reservoirs must not fail (criterion B). In terms of this present problem, these constraints can be represented mathematically as follows

$$x_1 = 80 \qquad x_2 + x_3 = 120 \qquad x_4 = 100 \text{ (criterion A)}$$
$$x_1 + x_2 \leqslant 250 \qquad x_3 + x_4 \leqslant 200 \qquad \text{(criterion B)}$$

Of the four unknown quantities (variables), it is observed that the values for x_1 and x_4 are fixed and only x_2 and x_3 can change. The problem of minimising the cost function

$$z = \sum_{i=1}^{4} c_i x_i$$

thus becomes one of minimising $c_2 x_2 + c_3 x_3$. If the costs are known, then the problem is solvable. For instance, assuming $c_2 = 1.0$ and $c_3 = 1.5$, then we must minimise

$$z = 1.0x_2 + 1.5x_3 \tag{6.1}$$

subject to the constraints (rewritten by substituting $x_1 = 80$, $x_4 = 100$)

$$x_2 + x_3 = 120 \tag{6.2}$$
$$x_2 \leqslant 170 \tag{6.3}$$
$$x_3 \leqslant 100 \tag{6.4}$$

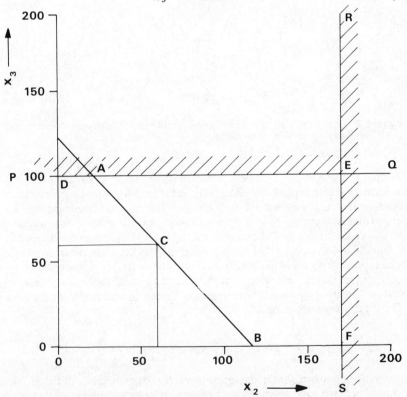

Figure 6.3 *Graphical representation for minimisation of cost function, z*

This set of equations can be solved using a graphical technique. In figure 6.3, the area below the line PQ and to the left of the line RS satisfies constraints 6.3 and 6.4 above. Since the flows must be forward and not backward, then the values of x_2 and x_3 must be positive. The viable region is thus further restricted to area ODEF. To satisfy constraint 6.2, then the solution must lie on the straight line represented by equation 6.2 and within the region ODEF. This line is the line ACB in figure 6.3. (C is the point at which $x_2 = x_3 = 60$.) The only possible solutions therefore lie on this line. Of the infinity of these solutions, which is the cheapest? By trial and error it can be calculated that at point A ($x_2 = 20, x_3 = 100$) the cost, z, is 170. At point C, $z = 150$ and, at point B, $z = 120$. It thus seems that, the further down the line ACB, the cheaper the solution becomes. The optimal solution must therefore occur at point B ($x_2 = 120, x_3 = 0$). It can be proved mathematically that this is indeed the cheapest solution, since equation 6.1 for z represents a set of parallel straight lines at a shallower angle than the line ACB; the intersection of any of these lines with ACB gives the cost directly (see figure 6.4).

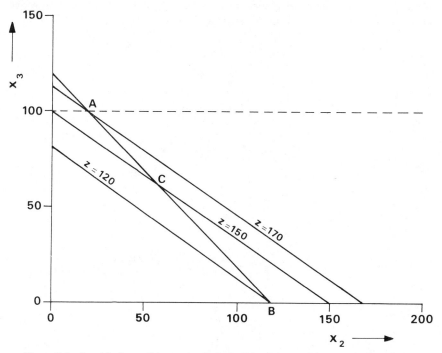

Figure 6.4 *Graphical proof that point B, derived by the use of figure 6.3, does indeed provide the minimum cost required*

A further advantage of this simple method is that it can also be undertaken using a matrix of numbers and a technique known as Dantzig's Simplex Method— a method eminently suited to rapid solution by computer.

If any further constraints are imposed on the system (for example, that pumping may only be done in integer units) the reference to the operational research literature provides many additional techniques (such as branch-and-bound techniques).

Local Distribution

Within this framework, the problem of local distribution—both spatially and temporally—can be discussed in detail. On leaving the water treatment works, the water will be transferred via a *trunk main* to the distribution area. In most cases, this main will deliver the water to a temporary storage area or *service reservoir* (which may be in the form of a *water tower*).

A service reservoir fulfills three main objectives.

(1) It provides a 'head' of water so that there is sufficient water pressure in the mains which deliver the water to the customer. Accordingly, service reservoirs are built on comparatively high land within the urban area, or on artificial hills (stilts) as water towers.

(2) It balances the fluctuating demands experienced over a period of 24 hours against the constant input source.

(3) It acts as a safeguard to maintain supplies if the trunk main is damaged.

Daily variations are most easily seen in the case of the domestic demand, in which peaks occur at around noon and 6 p.m. (see figure 6.5). Over the whole day the

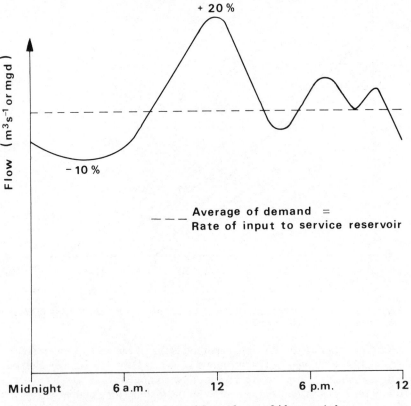

Figure 6.5 *Variation in total demand over a 24-hour period*

total input from the trunk main must exactly balance the total demand. During hours of peak demand, the supply rate to the consumer is greater than the supply rate to the service reservoir, and the volume of water stored there decreases. During the late evening and night, the demand falls below average and the excess (supply over demand) is retained by the reservoir as it is recharged. Such an arrangement enables diel fluctuations in demand to be satisfied without the necessity of using larger trunk mains.

Service reservoirs contain treated water and must therefore always be covered. In construction, they vary from reinforced concrete tanks with aluminium roofs to water bodies protected by a complete floating carpet of polystyrene spheres. Although the minimum capacity must be sufficient to satisfy the demands during one day, it is usual practice to have a capacity that can meet 3-day demands. This will cater for the weekly and seasonal fluctuations. The highest daily consumption used to occur on a Monday (washday), with a minimum on Sunday. Nowadays, with increased leisure use of water, Sunday is the peak demand day in many areas.

From the service reservoir, the water is fed into secondary mains of about 0.0100 to 0.0150 m in diameter (see figure 6.6), each of which serves an area of the town. These mains then subdivide into individual street mains (0.0075 to 0.0100 m), which serve the individual customers.

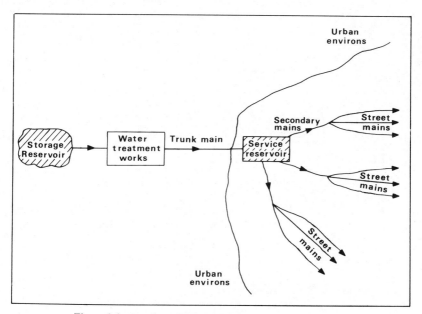

Figure 6.6 *Distribution of water from reservoir to consumer*

Summary

The problems of management (including distribution) have been discussed on differing time and space scales. On a macroscale, supplies to several townships from alternative sources (for example, storage reservoirs) may be optimised using operational research techniques, although accurate forecasts of demand are needed. Management and control are simplified by the increasing use of telemetry to convey information across the large distances between, for example river abstraction points and the pumped storage reservoir. On a mesoscale, service reservoirs are used to smooth the demands over periods of 24 hours, although the wise inclusion of safety margins in their designs permits them to balance the weekly and seasonal demand patterns.

Suggested Reading

Singh, J., *Operations Research* (Penguin, Harmondsworth, 1968).
Walsh, G. R., *An Introduction to Linear Programming* (Holt, Rinehart & Winston, London, 1971).

7

Temperature and Current Structure

To the engineer, the body of water in a reservoir may appear simply to be a stored supply of water that must be collected and redistributed as and when necessary. For many purposes this macroscale view of water is sufficient: it enables the engineer to plan the distribution networks, dams, water treatment works and so on that are necessary for the efficient management of a water resource, so as to meet domestic and industrial demands.

But this approach implicitly assumes one very important factor--that the bulk water is of a high enough quality to be useful. Water that has become tainted by algal growths or has been depleted of oxygen and is in a stinking, anaerobic condition is of little use. The determination and appreciation of water quality often falls heavily on the shoulders of the chemist, the biochemist or the biologist. The balance of nature in a closed (or almost closed) system such as a reservoir is often

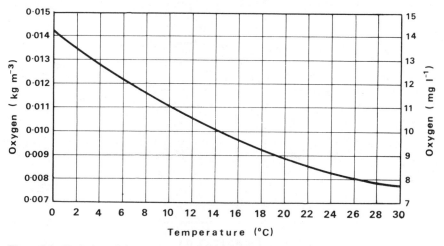

Figure 7.1 *Variation of the maximum value for the dissolved oxygen content of water as a function of temperature (reproduced, by permission, from Mills, 1972)*

delicate and unpredictable. Nutrient drainage from excessive use of fertilisers, for example, may easily upset the balance of the natural cycles resulting in eutrophication (see chapter 8 for further discussion).

Life that can exist in a reservoir is dependent on three main factors: light penetration, availability of oxygen and food supply. The last two of these will be discussed at length in later chapters. The amount of oxygen that can be dissolved in water, even at saturation, decreases markedly with increase in temperature (see figure 7.1) and its vertical mixing depends on the temperature gradient and currents in the reservoir. The oxygen content of the water and its temperature are among the factors that determine the palatability of the water. The quality of water withdrawn for potable use is thus strongly influenced by the temperature regime within the reservoir.

The Thermocline

Throughout the winter months most lakes in Great Britain (and elsewhere if they do not freeze over) exhibit an *isothermal* vertical profile. The water can then be considered as a single, well-mixed (*homogeneous*) layer at (usually) about 277 to 279 K (4 to 6 °C), and this will be the case until about the time of the vernal equinox (21 March) when the incoming solar radiation increases sufficiently to warm the reservoir.

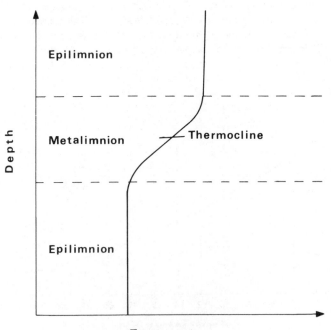

Figure 7.2 *Typical summer temperature profile in a reservoir*

As the heat begins to penetrate the lake, the upper layers warm first and a non-zero temperature gradient, dT/dz, arises; the temperature T is a function of the depth z and the time t. By mid-summer a temperature gradient as illustrated in figure 7.2 is observed in most lakes. In the first instance, we can consider the lake to be divided into two layers (or *stratified*), each of which is approximately isothermal: a warmer upper layer (the *epilimnion*) overlying a cooler lower layer (the *hypolimnion*). The interface is known as the *thermocline*, and this is where there is a rapid change of temperature between the two layers. The finite depth over which this steep temperature gradient exists is usually known as the *metalimnion*. Mathematically, the position of the thermocline is given when the second derivative of the temperature with respect to depth, d^2T/dz^2, is zero. Its very existence, characteristic of all deep water bodies (oceans as well as lakes and reservoirs), is of paramount importance in determining water quality. It presents a physical barrier between the two layers, severely inhibiting vertical downward mixing of momentum, heat, dissolved oxygen, etc. Without regular renewal of its supply of oxygen, the ability of the hypolimnion to support life diminishes during the period when the thermocline prevails. Anaerobic conditions arise and the water becomes of little use.

Thus a knowledge of temperature structure and its variation on both annual and diel time scales is important.

Annual Stratification Cycle

Water has its maximum density at 277 K (4 °C)—an important fact in lake stratification. If the reservoir is homogeneous at 277 K, as is often the case in mid-March, and heat is applied in the form of both short-wave solar radiation and long-wave atmospheric radiation, the lake water begins to warm. As the net radiation received increases, the thermocline forms and descends until summer stratification sets in (see figure 7.3). This continues until solar radiation decreases and the top layers of the lake begin to cool. At this stage the density profile is unstable, since the cooled water is more dense and starts to sink. Thus *convection* arises, and continues until a single homogeneous layer occurs again. This so-called *autumn overturn* is often assisted by the autumn equinoxial gales, strengthened by the energy input given up to the atmosphere by the oceans as they cool. As winter approaches, further cooling *may* occur but this time, due to the anomalous properties of water, the cooled water is less dense and so does not sink. This cool upper layer tends to be shallower than the epilimnion that formed in summer and may indeed cool sufficiently to freeze. Thus a lake freezes from the top downwards and the heat-transfer properties of ice must be considered in determining how the temperature of the water beneath the ice will behave. Lakes and reservoirs seldom freeze in Great Britain, partly because of the tendency of wind-induced waves to break up any ice that starts to form. As a frozen lake melts and warms, an unstable situation arises and the *spring overturn* occurs as the water temperature rises towards 277 K.

These two overturns assist in the transport both of essential nutrients from the base of the lake to the surface and of dissolved oxygen and heat down uniformly throughout the reservoir's depth.

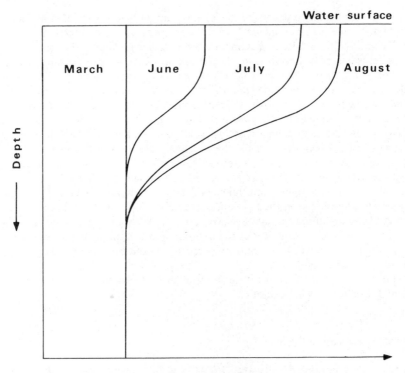

Figure 7.3 *Four typical temperature profiles illustrating the creation of an epilimnion (upper heated layer) during the summer months*

A lake that experiences two overturns each year—typical of temperate latitudes—is known as *dimictic*. If a lake is always ice-covered, however, no overturns occur and the lake is termed *amictic*. In certain regions the lake may undergo only one overturn. Polar lakes never reach a temperature of 277 K; thus they only experience the spring overturn and are *cold monomictic*. At the other extreme of climate, tropical lakes are always warmer than 277 K and are *warm monomictic*, experiencing only the autumn overturn. Many middle latitude lakes are also warm monomictic due to their location within a warm oceanic-type climate (for example, in Great Britain).

Diel Variation

So far we have considered the mean daily behaviour of the reservoir throughout the year. The thermocline that forms is *seasonal*. It is built up slowly over a period of days or weeks by the daily mixing down of heat to form a shallower thermocline each day, which disappears overnight—the *transient thermocline*. Figure 7.4 shows a transient thermocline typical of a day in July when a seasonal

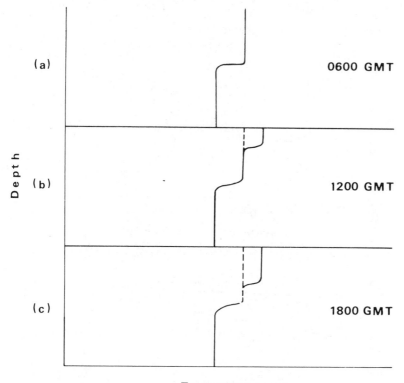

Figure 7.4 *The transient thermocline mimics the seasonal thermocline but on a shorter time scale; in the temperature profiles at three times of the day shown here, the transient thermocline is superimposed on a seasonal thermocline*

thermocline also exists. It can be seen from this diagram that the daily variation of the transient thermocline is analogous to the annual variation of the main seasonal thermocline. The main thermocline may or may not also exist, but as the heat is mixed down throughout the day the transient thermocline forms, superimposed on this structure. By late evening it may have dropped to the level of the main thermocline. If this happens the seasonal thermocline will be enhanced. Overnight cooling seldom removes as much heat as was put in during the day.

Temperature Structure

There are three main mechanisms that control the vertical distribution of heat in a reservoir

(1) differential absorption of light
(2) molecular and turbulent diffusion
(3) convection.

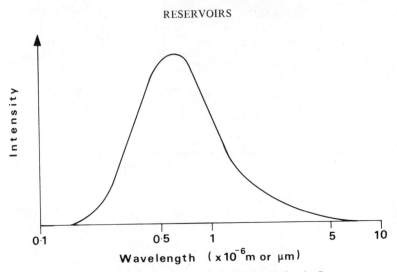

Figure 7.5 *Planck function (black-body curve) for the Sun*

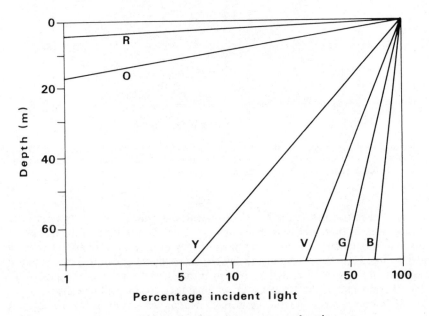

Figure 7.6 *Transmission of light by distilled water at six wavelengths*

$R = red$ $= 0.72 \times 10^{-6}$ m $= 0.72$ μm
$O = orange$ $= 0.62 \times 10^{-6}$ m $= 0.62$ μm
$Y = yellow$ $= 0.56 \times 10^{-6}$ m $= 0.56$ μm
$G = green$ $= 0.51 \times 10^{-6}$ m $= 0.51$ μm
$B = blue$ $= 0.46 \times 10^{-6}$ m $= 0.46$ μm
$V = violet$ $= 0.39 \times 10^{-6}$ m $= 0.39$ μm

The percentage of incident light that would remain after passing through the indicated depths of water is expressed on a logarithmic scale (after Wetzel, 1975, by permission of the author and publisher)

These mechanisms are each important at different times of the year and with different meteorological conditions. At any one time for a particular lake, one or more of the mechanisms may be dominant.

The strength or intensity of the sun's energy at different wavelengths is illustrated in figure 7.5—the *Planck curve* or *black-body spectrum*. It should be noted that the maximum strength is in the visible region—4.5 to 5.5×10^{-7} m (0.45 to 0.55 μm). This wavelength distribution of incoming solar radiation is important since the wavelength of light determines its penetrative power. Thus, blue light penetrates to a greater depth than red—all other things being equal (see figure 7.6). This solar or *short-wave* radiation is absorbed throughout the lake's depth, although most is absorbed near the surface, and only in very clear lakes (of low *turbidity*) does light penetrate to an appreciable depth. As it is absorbed the radiation heats up the surrounding water, and this absorption alone is capable of producing a temperature gradient in the lake. The penetrative ability of the light is often measured in terms of an *extinction coefficient* (evaluated either at an individual wavelength or, more usually, as a mean over the visible part of the spectrum). The value of this coefficient can vary from 0.026 m^{-1} (for pure water) upwards. (In turbid Lake Mendota in the United States it has a value of 0.88 m^{-1}). Its numerical value gives the depth at which the light has only (1/e)th of its surface value (e = 2.718282...). The extinction coefficient (and hence a measure of the

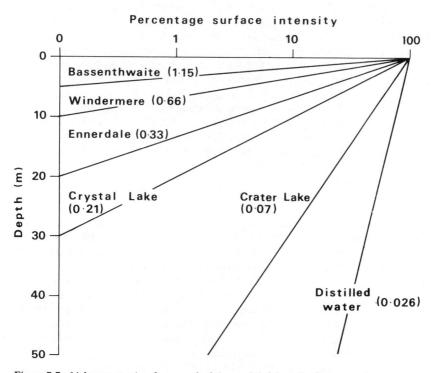

Figure 7.7 *Light penetration for several of the world's lakes; the figures on the curves are the respective values of the extinction coefficient, expressed in m^{-1}*

turbidity) are often deduced using a *Secchi disc*—a black and white quartered disc about 0.2 m (20 cm) in diameter. The depth at which this disc cannot be observed is inversely proportional to the extinction coefficient. The effect of turbidity on light penetration is illustrated in figure 7.7, where absorption curves are shown for different lakes in the world. The values for the extinction coefficient are also indicated in this diagram.

Diffusion of heat also occurs in a lake. It may be less or more important than light penetration. *Molecular diffusion* occurs all the time and is the process by which the molecules in the water transfer heat by virtue of their random motion. This random motion may be described by kinetic theory, from which it can be deduced that the rate of heat diffusion is proportional to the temperature gradient. The constant of proportionality is known as the *coefficient of molecular diffusion*, often denoted by α, and has a value of approximately 1.3×10^{-7} m^2/s, although this varies very slightly with the temperature of the water.

On the other hand *turbulent diffusion* may also exist in the lake as a result of large-scale eddy motions, which are set up in the water by currents. These *eddies* transfer whole 'lumps' of water vertically over a depth of up to several metres; any water brought down from above in this way brings with it heat. The currents themselves are often induced by a surface wind stress. It is found that this *eddy* or *turbulent diffusion* is again proportional to the temperature gradient but in this case the constant may be much larger, possibly up to 10^{-2} m^2/s, and is determined by the vertical velocity shear (that is, the variation of the horizontal current with depth). It is only on calm, nonturbulent days that eddy diffusion subsides sufficiently for molecular diffusion to play an important role in heat transfer.

The third process in which heat is transferred is *convection*. This is important as a mixing mechanism at certain times of the year only. When it does dominate the mixing process, it overturns the whole lake and mixes it thoroughly to homogeneity. Convection occurs when the lake becomes unstable: when the upper layers are denser than the lower layers, and sink. This can occur either when the temperature of the upper layer is less than that of the lower layer but above 277 K, or when its temperature is below 277 K but above that of the underlying water. These are the autumn and spring overturns, respectively, and may be assisted by wind action.

Surface Energy Budget

Most of the energy entering the lake does so via its surface and only short-wave radiation has any direct effect at depths below the surface. (Inflows and outflows do contribute to the energy budget, but are only significant for reservoirs with short retention time; heat losses through the sides and bottom of the reservoir can usually be shown to be negligible.) The net surface heat flow is the total energy available to be transferred downwards. This is the *surface energy budget*. When the budget is positive, more heat enters the reservoir than leaves it, the water warms and a thermocline may form. A negative budget occurs in autumn; the water cools and convectional mixing takes place.

The surface energy budget (S.E.B.) is complex and varies with time of day, time of year and many meteorological parameters. Figure 7.8 shows the main

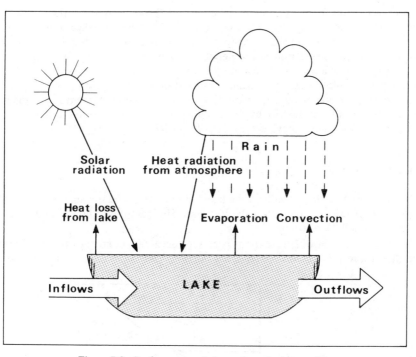

Figure 7.8 *Surface energy balance for a freshwater lake*

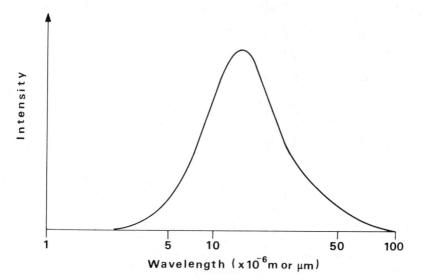

Figure 7.9 *Typical black-body spectrum for a lake surface*

sources and sinks of energy. Energy taken in and absorbed by the surface comes from the sun, the atmosphere and precipitation. It is lost because the water surface radiates away energy and also by evaporation. The long-wave energy is composed of electromagnetic radiation of wavelength between 4×10^{-6} and 10×10^{-6} m (4 to 10 μm) distributed as shown in figure 7.9. The energy is represented by the area under the black-body curve, and is proportional to the fourth power of the water-surface temperature. Any water that evaporates takes in latent heat, which is then removed from the lake by the escaping water vapour. Convection in the atmosphere may also remove energy (*sensible heat*).

Current Structure

We have seen how the currents in the lake help to determine the amount of turbulent diffusion and heat transfer in a lake. The question of how the current arises must now be answered.

Unlike a river, in which a current is present because its two ends are at different heights, a reservoir has, in general, a flat bottom. Two types of current can exist,

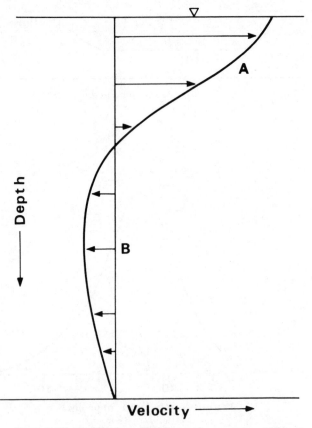

Figure 7.10 *Typical vertical structure of currents in mid-lake*

however. If a stream or man-made inflow feeds the reservoir then the incoming stream can create a current by simply 'pushing' the water already in the reservoir; this is known as the *hydraulically driven current.*

The effect of a wind (together with the Coriolis effect due to the rotation of the Earth—an important effect for large lakes) is to produce *wind-driven currents.* As the wind blows across the surface, frictional effects try to drag the surface along with the wind. A shear stress is set up at the water's surface and this is transmitted downwards. Unlike a solid, which can resist shear stress to some degree, the water cannot and is set in motion. At greater depths the effect of the surface stress decreases and the water moves less. Thus the wind-induced current is correspondingly smaller. It is immediately apparent that, in a closed basin, if the top water moves in one direction the lower water must move in exactly the opposite direction, or else there will be a huge 'pile-up' at one end of the lake. The surface current tends to be in the direction of the wind and gradually diminishes at lower levels. The *return current* is encountered at greater depths. If we consider a vertical cross section of the lake, the typical currents induced by the wind at the surface are of the form illustrated in figure 7.10. A plan view of a rectangular lake at the depths A and B (indicated on figure 7.10) is shown in figure 7.11.

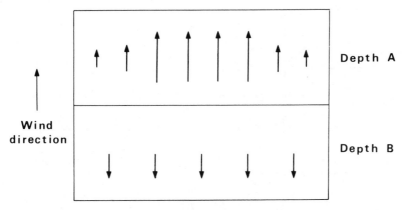

Figure 7.11 *Horizontal current directions at different depths*

Not only does the thermocline inhibit downward mixing across it (that is, downwards through it), as noted above, but it also provides a barrier for the viscous transmission of shear stress, and in a stratified (two-layer) lake the currents in the hypolimnion are found to be virtually zero, adding to the stagnation of these deep waters. The typical current structure shown in figure 7.10 becomes compressed into the epilimnion; the modified current profile for a stratified lake is shown in figure 7.12.

Thus in the epilimnion the relatively strong currents ensure that turbulent diffusion is the dominant heat transport mechanism. The much smaller velocities in the hypolimnion give rise to a coefficient of turbulent diffusion that is often much less than α, the coefficient of molecular diffusion, and hence molecular diffusion may be an important process in deep waters.

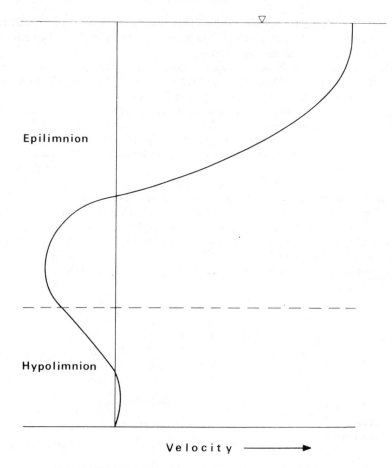

Figure 7.12 *Vertical variation of current in a stratified lake*

Summary

The temperature structure of the reservoir can be seen to have many important facets. It is directly involved in determining the temperature of the water that is withdrawn for use and indirectly involved in the quality of this water. The vertical temperature profile is found to depend upon the prevailing meteorology, especially the strength of the wind and the extent of the solar radiation. Downward heat transfer is the result of eddy motion set up by the wind-induced horizontal currents; the heat is released back to the atmosphere when the lake cools. The lake then becomes unstable, convection occurs and all the water in the lake is mixed to form an homogeneous body. These overturns are primarily responsible for the circulation of phytoplankton and nutrients and for the re-oxygenation process. The physics of the water body must therefore be considered in order to determine the biogeochemical cycles operating within the lake.

Suggested Reading

Hutchinson, G. E., *A Treatise on Limnology*, Vol. 1 (Wiley, New York, 1967).
Wetzel, R. G., *Limnology* (Saunders, Philadelphia, 1975).

8

Biochemical Cycles and the Quality of Stored Water

Tap water is not chemically pure hydrogen oxide (H_2O): if it were, then the public would, in general, find it unpalatable. It also contains traces of various chemicals, which, occurring naturally in the water, give it what is regarded as an acceptable 'taste'.

In a lake or reservoir, the growth and decay of the biota are determined by the chemical and physical properties of water. This chapter considers the interactions between biological and chemical cycles present in the reservoir. The advantages and disadvantages of the ways in which the cycles determine the biological and chemical content of the water are discussed, with special reference to the resulting water quality.

Chemical Properties of Water

Water is a fascinating chemical compound. A water molecule contains two hydrogen atoms and one oxygen atom, joined at an angle of $105°$ (see figure 8.1). The electrical charge on the molecule tends to separate: the oxygen acquires a net negative charge and the hydrogen a positive charge, thereby creating a *dipole*. Two or more molecules of water are thus held together by an electrostatic attraction (oxygen to hydrogen—see figure 8.2). This *polymerisation* is responsible for many of the observed chemical properties of water.

Some of the molecules *ionise* completely, splitting up into free H^+ and OH^- ions. The concentration of the H^+ ions is a measure of *acidity*, expressed using the *pH scale*, which usually runs from 0 to 14. At low pH values, acid conditions exist but as the pH value increases, the conditions become more *alkaline*. The transition point between acid and alkali occurs at a pH of 7.0—the pH of pure water, which is thus *neutral*.

Although it is neutral (and thus noncorrosive), water is a *solvent* and will dissolve almost any other chemical to some degree. From a water quality viewpoint some of these chemicals are more important (and more common) than others.

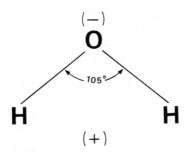

Figure 8.1 *Chemical structure of a water molecule*

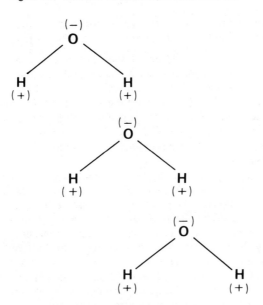

Figure 8.2 *Polymerised structure of water*

At least 19 elements (all of which can be found dissolved in water) are essential for life (see table 8.1), although some of these are only needed in minute traces. These basic elements are known collectively as *nutrients*. Three of these demand detailed consideration: oxygen, nitrogen and phosphorus.

The Role of Oxygen

The maximum amount of oxygen that water can contain is, in theory, the amount needed to make the water *saturated*. However, this value depends critically on the water temperature (see figure 7.1). At an average water temperature of 283 K (10 °C), 0.011 kg/m^3 (11 mg/litre) of oxygen are needed to saturate the water, whereas water saturated at a temperature of 291 K (18 °C) will contain only 0.009 kg/m^3.

Table 8.1   Essential Nutrients for Plant Growth

Element	Boron	Carbon	Calcium	Chlorine	Cobalt	Copper	Iron
Chemical symbol	B	C	Ca	Cl	Co	Cu	Fe

Element	Hydrogen	Potassium	Magnesium	Manganese	Molybdenum
Chemical symbol	H	K	Mg	Mn	Mo

Element	Nitrogen	Sodium	Oxygen	Phosphorus	Sulphur	Vanadium	Zinc
Chemical symbol	N	Na	O	P	S	V	Zn

The amount of oxygen present in the water can be measured in two ways:

(1) as the total quantity present (a measure independent of temperature);
(2) as the degree of saturation (expressed as a percentage).

Although the dissolved oxygen (D.O.) content is often expressed in terms of a saturation value, it is the actual amount present that determines the possible biota. Larger animals (such as fish) require, typically, 0.008 kg/m^3 of oxygen for active life. Below about 0.004 kg/m^3, fish die and as the D.O. content decreases further only bacteria and other saprobic organisms remain. If the D.O. level falls to zero, then the lake becomes *anoxic*. If, at the same time, nutrients remain in the water then *anaerobic* conditions exist and only a limited number of types of bacteria can thrive.

Depending on the oxygen content, two different classes of bacteria may exist in freshwater. If oxygen is present, *aerobic* bacteria use it to respire and to oxidise nutrients, releasing carbon dioxide, water, etc. These die as the free oxygen is depleted and are replaced by *anaerobic* bacteria, which no longer require gaseous oxygen in order to oxidise the chemical compounds present in the water. The gases liberated by these bacteria are hydrogen sulphide (H_2S), methane (CH_4) and ammonia (NH_3)—all obnoxious and evil-smelling, typical of stagnant or organically overloaded waters. Anoxic conditions are undesirable in an aquatic environment whose water is destined for human consumption. Reservoirs may become anaerobic at certain depths and at certain times of the year.

Oxygen is continually being used in respiration by the animal and plant life at all depths. Photosynthetic plants produce oxygen as a by-product, but these are confined to the upper levels of the lake where light can penetrate. This is the *euphotic zone*. An imbalance may occur in the distribution of the sources (photosynthetic plants) and sinks (respiring biota) of dissolved oxygen. (Photosynthesis and plant and animal growth are discussed in more detail in chapter 9.)

At the free water surface, oxygen will diffuse into the water from the atmosphere above (unless the top layer is already saturated or supersaturated). The gas molecules are then transported downwards by molecular diffusion, eddy diffusion and convective currents. These transfer processes operate in the same way and in a similar annual cycle to the corresponding processes for heat transfer described in chapter 7, and help to maintain the supply of oxygen throughout the lake.

If summer stratification occurs, vertical transport through the thermocline is severely restricted and the oxygen levels in the hypolimnion can no longer be replenished. The aerobic populations (animals, plants, bacteria) utilise all the available oxygen and then die, providing food for the anaerobic decomposers. It is for this reason that major research is being undertaken to find ways of achieving artificial *destratification* and maintaining an homogeneous reservoir throughout the year.

These themes are best illustrated by graphs showing *isopleths* (lines joining points of equal magnitude) of dissolved oxygen. In figure 8.3 the D.O. content is plotted as a function of both lake depth and time of year. In January, the oxygen level at any depth is about 70 per cent of the saturation value. As the summer progresses, D.O. values in the hypolimnion decrease to less than 50 per cent, but the surface layers remain saturated (or even slightly supersaturated if the photosynthetic plants are abundant). The autumn overturn in September redistributes the oxygen equally to all depths. If the reservoir were to remain stratified, water withdrawn from the hypolimnion would contain very little dissolved oxygen and could contain appreciable concentrations of methane, hydrogen sulphide, detritus and nutrients (discussed in detail later in this chapter). Withdrawals for public supply would have to be made from the epilimnion. To avoid this restriction some method of artificially enhancing the D.O. content of the lower depths of the reservoir must be introduced. This may be of two basic types: destratification and re-aeration.

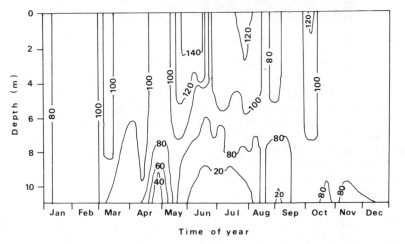

Figure 8.3 *Variations in the D.O. content of Farmoor Reservoir, near Oxford during 1968; the isopleth values are in terms of percentage saturation (after Youngman, 1975, by permission of the author)*

Destratification and Re-aeration

Destratification is the destruction of the two-layer structure of a lake that develops naturally during the summer months. It can be accomplished in many ways and is a means of totally mixing the two layers so that the temperature is constant and the D.O. content uniform throughout the reservoir. With some methods of destratification, this means that the D.O. content of the upper parts of the lake is lowered since no extra oxygen is introduced; the existing oxygen is simply redistributed. One such method is *jetting*, employed in many reservoirs in Europe (see figure 8.4). This is useful in a pumped storage scheme. The incoming water is pumped into the reservoir at a higher velocity than would otherwise be necessary. (This is easily accomplished by decreasing the diameter of the inflow pipe—the *venturi effect*.) The high-velocity stream of water forms a *jet*, which is responsible for introducing a large degree of turbulent mixing into the lake, usually in the hypolimnion, and destroying the thermocline.

Figure 8.4 *Jet installed in Farmoor Reservoir (courtesy of Thames Water—Vales Division)*

Since jetting results in an isothermal reservoir, the quality of the water withdrawn becomes independent of the depth from which it is taken: the whole lake has an acceptable oxygen content. Since it has been artificially mixed its temperature is a little lower than the previous epilimnion value. In temperate climates, such as the British Isles, this will be of the order of 288 to 293 K (15 to 20 °C), but in lower latitudes (for example, the mid and southern United States) the isothermal temperature that results from jetting may be as high as 303 K (30 °C). Water at such high temperatures is not in general acceptable to the public as a potable supply. In these areas, the temperature typical of hypolimnetic water from a stratified lake would give a more pleasant 'taste' than the water from a well-mixed lake—the temperature of the water from the hypolimnion of a low-latitude stratified lake is often about 283 to 288 K (10 to 15 °C)—but the oxygen content of this water is too low. To overcome this problem, many American lakes use a system of *hypolimnetic re-aeration*, in which the stratified nature of the lake is maintained, but the D.O. content of the hypolimnion is enhanced artificially. This is done by introducing compressed air or oxygen in the lower depths of the lake, at a very low pressure. Care must be taken that the thermocline remains intact, since one method of destratification relies on the release of high-pressure gas bubbles in the hypolimnion: the bubbles destroy the thermocline as they rise rapidly upwards, thoroughly mixing the water in the process.

Nutrients

Plant growth ceases if *any one* of the essential nutrients listed in table 8.1 is missing. Such a shortage is a *limiting factor* in plant growth. These nutrients are usually available in the minute quantities required. Five of them, however, are required in larger amounts: carbon, hydrogen, oxygen, nitrogen and phosphorus. These are needed in proportions of approximately $106:181:45:16:1$. The first three (C, H, O) are readily available either in water (H_2O) plus dissolved carbon dioxide (CO_2) and thus never become a limiting factor in aquatic plant growth. However, the concentrations of nitrogen and phosphorus dissolved in natural waters are much lower and it is nearly always one of these elements that provides the limiting factor for growth. The relative importance of these two for lentic freshwater bodies is still in dispute. Recent investigations seem to favour phosphorus as the controlling factor, although it may be that different species respond to different nutrients and for a stable multispecies situation each species is controlled by a different nutrient.

Nitrogen and Phosphorus

Nitrogen and phosphorus are two vital nutrients for plant growth. Some plants, such as the blue-green algae, are capable of acquiring nitrogen direct from the atmosphere by *fixation*, but most plants need to obtain their nitrogen intake in the form of nitrogen compounds, such as nitrates, found in the water. Similarly, phosphorus is often utilised in the form of phosphates or orthophosphates. Excess nutrients can lead to prolific plant growth. Rapid reproduction of plants such as pondweed and algae can result in undesirable conditions in the lake water. *Algal blooms* (see chapter 9) often add taste and colour to the water. As little as 10^{-5} kg/m^3 (0.01 mg/litre) of phosphorus can trigger such a bloom. In 1968 Lake Erie already contained an average of 4×10^{-5} kg/m^3 (0.04 mg/litre). If the phosphorus level rises to 10^{-4} kg/m^3 (0.1 mg/litre) and the nitrogen level to 3×10^{-4} kg/m^3 (0.3 mg/litre) at the time of the spring overturn, then an algal bloom during the summer is highly probable. The nutrient loading may also be responsible for blooms even when the mean concentrations are relatively low (for example, Esthwaite Water has nutrient levels lower than 10^{-5} kg/m^3 but still has regular algal blooms). Nitrates and (ortho)phosphates occur naturally, but can also be added to lake water as a result of the excessive use of these compounds in agricultural fertilizers. Used on the land, they are eventually *leached* into watercourses, and thence lakes, and enhance the nutrient level. (The phosphorus loading from this source is low as a result of absorption by the soil.) Detergents and sewage add to this effluent load; indeed domestic sewage is probably the major cause of enrichment in many lakes. Manufacturers of detergent have already taken major steps to limit their contribution to the phosphorus loading of lakes; initially, detergents contained 35 to 50 per cent of sodium tripolyphosphate ($Na_5P_3O_{10}$), which resulted in millions of tonnes (1 tonne = 1000 kg) of phosphate being added globally to waste water each year. Increasing the efficiency of sewage treatment and banning phosphate-containing detergents are two control measures that have already been applied successfully in some countries. A completely satisfactory alternative to phosphate in detergents does not yet seem to have been found and other compounds (such as nitrilotriacetate, NTA) have been tried with limited success.

It is technologically possible to reduce the plant nutrient levels in a lake simply by controlling the amount of nitrogen- and phosphorus-enriched compounds used in agriculture and by controlling the way in which they are used by consideration of drainage patterns, collection and purification schemes, etc. If phosphorus is indeed a major limiting factor for growth (cf. nitrogen) then only phosphate use needs to be limited. This is less difficult than restricting nitrogen input to a lake both because of the habitual usage of nitrates and also because direct nitrogen input from the atmosphere may be an important (natural) source of nitrates in the lake.

Eutrophication and Some Remedies

Lakes and reservoirs can be classified in terms of their productivity—and thus their nutrient loadings. The classification is often linear in time. Young lakes will usually have low nutrient content, little decaying matter and high dissolved oxygen levels; such lakes are *oligotrophic*. They are normally deep and have cold hypolimnions. Algal blooms are rare occurrences, although the number of species of algae present may be large.

As a lake ages, silt accumulates on the bottom. In the *littoral zone*, more plants are able to take root, and the surface area of the lake that is free from rooted vegetation slowly decreases. At the same time, abundant growth and a continuous inflow of nutrients increase the productivity of the lake. There is a large amount of suspended solids and the hypolimnion contains little or no dissolved oxygen; algal blooms are commonplace. The lake is termed *eutrophic*.

As the lake silts up and turns into a marshy waste, the lake may be termed *dystrophic*: it contains large amounts of humic material, but little oxygen or nutrients and only few algal species. However, if leaching of nutrients from the land to the lake has been rapid, few nutrients enter the ageing lake. Such lakes will revert to being oligotrophic in old age.

Although lake *eutrophication* can be a natural process, its artificial acceleration is causing concern throughout the world. Through the recycling of the nutrients, the algae flourish and degrade the purity of the water. Nutrients are taken up by the algae and released during the anaerobic decomposition of dead algae, which takes place on the lake floor; the released nutrients are then brought back to the surface by the overturns, and so they become available (together with the new nutrient inflows) for another crop of algae. The availability of both nutrients and light is important for algal growth. Nutrient recycling occurs as a result of autumnal overturning and also within the epilimnion.

In a eutrophic lake, algae can be a severe nuisance. Some species add colour and taste to the water; some species release toxic substances; and they can all be responsible for the clogging of filters at withdrawal points. Algae and plants may be controlled by chemical, mechanical or biological means. Copper sulphate is often used as an algicide. Aquatic weeds may be treated similarly, or with some of the chlorinated benzenes and arsenical compounds. However, great care must be taken in their application, so as to avoid any further interference with the ecological balance locally. Larger weeds may be dredged or cut, although this is only a temporary measure. In Africa, Asia and the southern United States, the water hyacinth (*Eichhornia*) is a major problem; under the right conditions, this weed can double in number every 8 to 10 days. Treatment with sodium arsenite was used until it was realised that this was a hazard to human health. Mechanical harvesting and hormone treatment have also been tried but with only limited success. Surprisingly, recent work indicates that the weed may provide a useful source of energy,[1] livestock feed and fertiliser and can be useful in the treatment of sewage, by extracting many of the toxic heavy metals.

The biological control of weeds has often been initiated in the United Kingdom, for example, by the Anglian and Thames Water Authorities. Species of grass carp (some of Chinese origin) have been introduced into the rivers and reservoirs in an attempt to eliminate many of the plants and water weeds that are abundant in these waters.

Summary

Maintaining high dissolved oxygen levels is important for good-quality water. New techniques of measuring chlorophyll *a* values *in situ* (for example, using fluorimetry) will assist in determining accurate values of dissolved oxygen and are currently being developed in order to provide a fuller understanding of the phenomena described in this chapter. Stratification may lead to an anaerobic hypolimnion, which can be avoided or rectified either by initiating an artificial overturn in a stratified lake or by augmenting the oxygen levels in the deep water below the thermocline. Nutrient levels determine the plant growth, which can be limited by a shortage of any one of 19 elements (although it is found that the limiting factor is usually phosphorus or nitrogen). Natural ageing may be accelerated by an inflow of excess nutrients. Under these conditions the reservoir may rapidly become eutrophic and of less use for water supply.

Suggested Reading

Fair, G. M., Geyer, J. C., and Okun, D. A., *Water and Wastewater Engineering*, Vol. 2 (Wiley, New York, 1968).

Reid, G. K., and Wood, R. D., *Ecology of Inland Waters and Estuaries* (Van Nostrand, New York, 1976).

Wetzel, R. G., *Limnology* (Saunders, Philadelphia, 1975).

---

[1] When dried and burnt as a fuel, water hyacinths have a calorific value about two-thirds that of methane.

9

Freshwater Ecology

In most reservoirs, there exists a plethora of living organisms, ranging from viruses and bacteria to fish and even mammals and birds. These interact in complex and diverse ways both with each other and with the geochemical cycles discussed previously. Some biota, such as algae, bacteria and viruses, may affect the water quality directly causing a health hazard; others may simply be a prerequisite for the successful functioning of the ecosystem.

Ecosystems

In a natural environment interactions between the many organic and nonorganic components can be described in terms of an *ecosystem* (see figure 9.1). The components of an ecosystem can be classified into four groups

(1) abiotic components, such as nutrients
(2) producers, such as photosynthetic plants
(3) consumers, such as fish
(4) decomposers, such as bacteria.

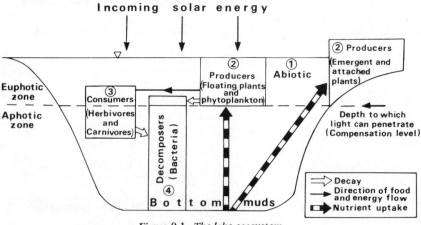

Figure 9.1 *The lake ecosystem*

80

Energy flows through this ecosystem, entering in the form of solar radiation. This energy is fixed by the process of *photosynthesis* by the *primary producers,* which are sometimes known as *autotrophs* since they can obtain nourishment directly from inorganic material present in the water. Photosynthesis can be described by the equation

$$6H_2O + 6CO_2 \xrightarrow{\text{light}} C_6H_{12}O_6 + 6O_2$$

Organisms that need to ingest other plants or animals are known as *hetero-trophs*: these are the system's *consumers*. In a reservoir this group ranges from minute zooplankton through herbivores to larger carnivores. Another type of organism that would fit into this category is the bacterium. However, *micro-consumers* (which include bacteria) play such an important role in the breakdown of decaying organic matter that they are classified separately as *decomposers* or *saprophytes*. Typical of the inorganic compounds released as by-products are the nutrients, which are eventually recycled by the physical and chemical processes in the reservoir and are thus made available again to the primary producers.

Figure 9.1 also shows the direction of food flow within the ecosystem. There are many ways of investigating this flow: in terms of food mass, an energy flow, a food chain, a pyramid or a complex food web. The number of steps along a food chain helps to define the organism's *trophic level*. Higher organisms belong to higher trophic levels.

The abiotic components of the ecosystem interact with the life forms in many ways. The origin of life on the Earth most probably occurred in an aquatic environment. One of the many reasons for this was the ability of the water to support large masses by virtue of buoyancy, enabling animals to develop without the need of supportive organs. Temperatures in water have smaller annual and diel ranges than in the atmosphere, because of the relatively large thermal capacity of water. Metabolic rates depend on temperature and many animals can only survive within a limited temperature range. This temperature moderation assists aquatic animals to thrive and permits a large diversity of animals to coexist.

The light climate is especially important to the photosynthetic plants in the reservoir. The depth of light penetration depends on the turbidity of the water and on the presence of dissolved substances that absorb and scatter light. Although this depth may change during the year, the annual average depth to which light penetrates can be deduced for a given lake. This is known as the *compensation level* (see figure 9.1) and is the level above which, in the *euphotic zone*, energy fixed by photosynthesis exceeds energy used in respiration and thus the plant can grow. Below the compensation level, the light available is insufficient for plants to thrive, and thus there are few producers to be found in this *aphotic zone*.

Fish and Other Consumers

The most common representatives of the higher trophic levels are fish, although fish-eating mammals (such as otters) and birds may sometimes feed from the lake. The species of fish present in a body of water depend, in part, on the degree of eutrophication. Fish that prefer oligotrophic lakes with low turbidity, high oxygen content and cold water include many of the more highly prized fish such as trout, grayling and salmon. More tolerant fish can withstand warmer water, higher turbidity and lower oxygen levels. Many of these belong to the carp family, such as roach, bream and carp, and are the mainstay of angling in these lakes.

Smaller consumers include herbivores and carnivores. Most of the insect life of lakes is typical of any body of freshwater (that is, rivers and ponds)—water beetles, water boatmen, larvae, etc. There may also be specimens of mayfly nymphs (for example, baetis) and chironomid larvae (see figure 9.2). Worms and snails are likely to be present. Smaller, often microscopic, animals (zooplankton) feed on the phytoplankton (microscopic plants) and in themselves provide food for the larger carnivores; copepods (for example, *Cyclops*), cladocera (for example,

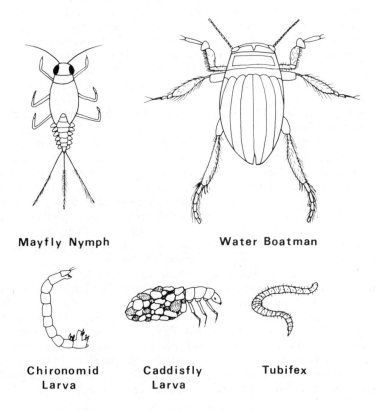

Mayfly Nymph   Water Boatman

Chironomid   Caddisfly   Tubifex
Larva   Larva

Figure 9.2 *Insects and animals found in reservoirs*

Daphnia) and rotifers (for example, *Philodina*)—see figure 9.3—are representative and are often present in large numbers. *Protozoa* are simple, single-celled zooplankton, 10 to 100 × 10^{-6} m long, that occur in freshwater. These animals, which include *Amoeba, Rhizopoda* and *ciliates* (see figure 9.4), feed mainly on bacteria (discussed later in this chapter).

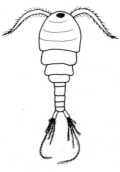

DAPHNIA   **COPEPOD**   **ROTIFER**
                      e.g. Cyclops   e.g. Philodina

Figure 9.3 *Typical zooplankton found in reservoirs*

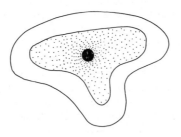

Amoeba

Ciliate   **Rhizopoda**
e.g. Paramecium   e.g. Arcella

Figure 9.4 *Typical protozoa found in reservoirs*

The phytoplankton on which zooplankton feed are small plant-like organisms that float at different levels in the water; these are the primary producers in the system.

Photosynthetic Plants

Photosynthetic plants occur in many sizes and to several degrees of complexity— from large water plants such as rushes and crowfoot to the microscopic algae. Rooted plants grow around the periphery of the lake; this area is known as the *littoral zone*. If the water is too deep for rooted plants then only floating plants and phytoplankton can thrive. Typical floating plants in the limnetic zone include many flowering species such as hornwort (*Ceratophyllum*) and duckweed (*Lemna*). Some flowering plants, whose leaves float partially submerged, are rooted in the bottom muds; examples include water milfoil (*Myriophyllum*), amphibious bistort (*Polygonum amphibium*), water crowfoot (*Ranunculus aquatilis*), yellow water lily (*Nuphar lutea*) and Canadian water weed (*Elodea*). Some rooted plants have erect aerial leaves. These are the reeds, which colonise the edges of reservoirs and infiltrate as the silt level builds up.

The smaller and flowerless photosynthetic plants are probably of greater interest in reservoir management. The most important of these belong to the group encompassing the *algae*. There are four main types of algae, three of which are classified in terms of their colour: blue-green algae, green algae, yellow-brown algae and the diatoms.

The blue-green algae (or *Cyanophyceae*) are the simplest. They occur as single cells or in filaments or else as a gelatinous mass, which may be visible to the naked eye. Algae are autotrophic and some blue-green algae have the additional ability to fix atmospheric nitrogen when dissolved nitrogen salts are unavailable. Optimum conditions for growth usually occur at some time during the summer period. In nutrient-rich waters, large algal biomass may be produced. The masses of algae can be clearly visible as a green scum on the water surface (or sometimes just below the surface, as in the case of dinoflagellate *Ceratium* in Esthwaite Water), which is known as an *algal bloom*. Such a bloom may necessitate the withdrawal of the reservoir from the operating system, since algae can both block off-take filters and also, in some species, release toxic chemicals into the water. Growth ceases when the nutrients are completely utilised. Often the growth is so great that self-shading occurs and the photosynthetic process is prevented at lower levels.

Important odour- and taste-producing blue-green algae include *Anabaena, Aphanizomenon, Gomphosphaeria, Microcystis* and *Oscillatoria* (see figure 9.5). *Anabaena* and *Microcystis* have been implicated in cattle deaths.

Diatoms, which belong to a second important group of freshwater algae (*Bacillariophyceae*), may add tastes to the water and often a fishy odour. These algae are yellow-green or brown in colour. In diatoms the cell wall is silica based. Common diatoms are *Asterionella, Fragilaria, Melosira* and *Tabellaria*. These are illustrated in figure 9.6.

In the third algal group (the green algae or *Chlorophyceae*), the cell structure is quite well defined. Some genera possess flagella and are motile. (A flagellum is a whip-like protrusion providing motility.) Fishy odours may be associated with their presence. Some may be colonial and form large aggregates of individuals.

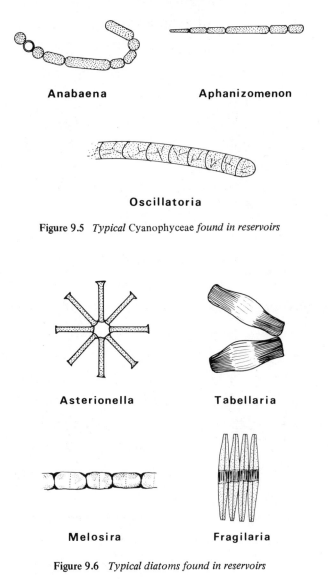

Anabaena   **Aphanizomenon**

Oscillatoria

Figure 9.5 *Typical* Cyanophyceae *found in reservoirs*

Asterionella   **Tabellaria**

Melosira   **Fragilaria**

Figure 9.6 *Typical diatoms found in reservoirs*

These are often visible as unsightly clumps drifting in the reservoir.

The final group are the yellow-brown algae (*Chrysophyceae*). Some types are multicellular and possess flagella. Their odorous releases may be very noticeable; for example, *Synura* imparts a bitter, cucumber-like taste to the water.

These are the four largest groups represented in a reservoir. One smaller group worthy of mention is *Dinophyceae* (dinoflagellates). A common dinoflagellate in many English lakes and reservoirs is *Ceratium*, which is responsible for blooms in certain lakes. These last three groups are represented in figure 9.7.

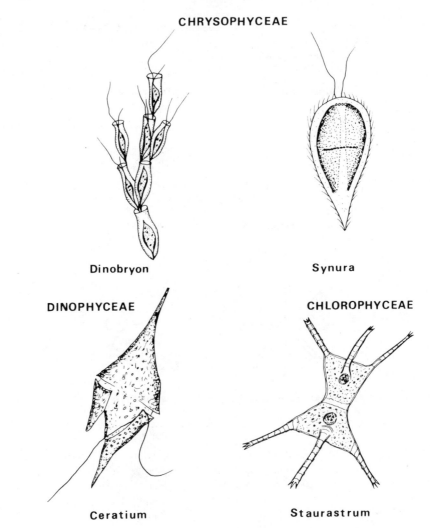

CHRYSOPHYCEAE

Dinobryon

Synura

DINOPHYCEAE

CHLOROPHYCEAE

Ceratium

Staurastrum

Figure 9.7 *Typical algal types* (Chrysophyceae, Dinophyceae, Chlorophyceae) *found in reservoirs*

Bacteria

The *bacteria* constitute an extremely important group of nonphotosynthetic organisms. They hold a unique role as *decomposers*: some (*aerobes*) utilise free oxygen to break down the organic matter of dead plants and animals, while others (*anaerobes*) use oxygen found in salts for the same purpose. Microscopic in size (from 0.5 to 5×10^{-6} m) they exist as spheres (cocci), rods (bacilli) or spirals (vibrio or spirilla); see figure 9.8. Reproduction is by fission and may occur, under

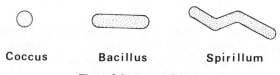

Coccus   **Bacillus**   **Spirillum**

Figure 9.8 *Bacterial types*

the correct physical conditions, every 15 to 30 minutes. Oxidation of organics is thus rapid, releasing carbon dioxide, water and energy. However, a second attribute of some bacteria is also worthy of detailed consideration. Several bacterial types are responsible for disease in humans; these are known as *pathogens*. Three common water-borne bacterial diseases (all transmitted via infected human faeces) are typhoid (caused by the bacterium *Salmonella typhi*), Asiatic cholera (*Vibrio comma*) and bacillary dysentery (*Shigella dysenteriae*). Such diseases have been virtually eliminated in developed countries by *efficient water quality control*,[1] although they are still endemic in some countries of the Third World. In these countries there is often little discrimination between water used as a supply and as a waste-disposal system. The absence of treatment either of incoming water or of sewage effluent allows the rapid infection of a community. Many pathogenic bacteria are easily removed (see chapter 12). Cholera can be virtually eliminated by the installation of slow sand filters.

To determine the efficiency of a selected treatment process, evaluation of the many biochemical interactions might seem necessary but impossible due to their complexity. It is however fortuitous that there is one bacterium, *Escherichia coli* (or *E. coli*) which is not only commonly found in large numbers in the intestinal tract of Man, but is also more resistant to standard water treatment than most pathogens. The bacterium itself usually produces no ill effects in humans and so is useful as an *indicator organism*. Testing water for the absence of *E. coli* will indicate (to a large percentage probability) that other bacteria (such as *Salmonella*) are not present. This is found to be an effective form of assessment of the success of water treatment methods.

The harmful effect of bacteria depends to a large extent on the number ingested. A second group of pathogenic organisms, *viruses*, are harder to control. These occur in much smaller numbers than bacteria (for example, 100 per litre of polluted Thames river water) and thus sample testing becomes less reliable. Viruses are smaller than bacteria (less than 0.3×10^{-6} m) and can only be seen using an electron microscope. Routine tests involve the injection of cell culture plates and Cytopathic Effect (C.P.E.) counts. Viruses can only reproduce within a specific host cell and are thus parasites. However, once reproduction becomes possible, their virulent effects become apparent. In Man, viruses are responsible for poliomyelitis and infectitious hepatitis. Unfortunately *E. coli* cannot be used as an indicator here, since viruses are much more resistant to water treatment than *E. coli* (by a factor of 40 in the case of the polio virus, for example). A further complication is that one viral type may cause quite different diseases in different people and dissimilar viruses often produce almost identical symptoms.

---

[1] The last outbreak of typhoid in Great Britain was in Croydon in 1937, when 43 people died. The Aberdeen typhoid scare in 1964 was not a consequence of water quality control mistakes in Great Britain.

Water quality

There are other, nonbacterial diseases associated with water. Certain diseases may
be caused by types of fluke. One of these is responsible for schistosomiasis (or
bilharzia), prevalent in certain Middle Eastern countries. The fluke spends part of
its life cycle in Man, in the bladder or intestine. Here it produces eggs, which pass
out of the body in the human excreta and often find their way into the irrigation
ditches in an area such as Aswan. To develop, the eggs must be ingested by a
specific host: a genus of freshwater snail indigenous to these ditches. The larval
flukes can then bore through skin to return to the human body and mature, thus
completing their life cycle. Another common disease transmitted, at some stage,
with the help of water is amoebic dysentery. This is caused by a protozoon
(*Entamoeba histolytica*) that is found in freshwater during part of its life cycle.

The biologically determined water quality is assessed in terms of the number of
pathogenic organisms present, taking into account whether these are natural or
artificially introduced. The number and type of algae present may be important
in determining the type of water treatment required. Some organisms are of course
desirable; for example, aerobic bacteria decompose unwanted organic material.
Knowledge of the aquatic biology and the interactions between the various life
forms permits the reservoir manager to optimise operating conditions. In a simple
reservoir with only one draw-off point, this may result in a decision to remove the
reservoir from the supply network during an algal bloom. Larger reservoirs are
often more adaptable. The shape of Rutland Water (see figure 9.9) is such that

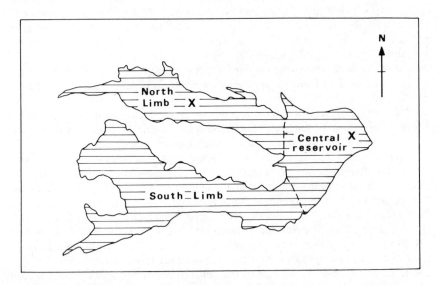

Figure 9.9 *Plan of Rutland Water; the two withdrawal points are indicated by crosses*

there are three distinct regions: a north limb, a south limb and the main central
part (the deepest). Scaled-down physical modelling shows that the interchange
of water between these three regions is minimal and water circulates within each

area. Thus if an algal bloom in one limb causes the quality to deteriorate there, only that limb needs to be removed from service and water can continue to be withdrawn for use from the (untainted) remainder of the reservoir. Two separate withdrawal points have been installed in the completed reservoir (see figure 9.9).

Summary

The biotic component of the ecosystem in a reservoir is important in determining the over-all water quality. Ill effects are likely to result from the ingestion of any pathogenic organisms that may be present in the water body. The ecosystem is divided into producers, consumers and decomposers, together with the necessary nutrients (abiotic component). Producers utilise sunlight to produce carbohydrates by photosynthesis. These are consumed by herbivores (zooplankton and higher animals) in the next step of the food chain. Food and energy are transported up the trophic pyramid to fish, mammals and Man.

A detailed knowledge of the type of algae (phytoplankton), bacteria and consumers at all trophic levels permits some control to be exercised in making water available for treatment and consumption. Undesirable algal blooms may cause less of a problem in large lakes, where they can be confined within a small percentage surface area of the lake. If allowed to persist, a large algal biomass may easily block filters at draw-off points and foul pumps, thus rendering the reservoir system inoperative and the water of low quality.

Suggested Reading

Belcher, H., and Swale, E., *A Beginner's Guide to Freshwater Algae* (H.M.S.O., London, 1976).

Mills, D. H., *An Introduction to Freshwater Ecology* (Oliver & Boyd, Edinburgh, 1972).

10
Numerical Modelling

Numerical modelling is an investigative tool which is rapidly gaining importance with the advent of bigger and faster digital computers. Any *model* is an accurately scaled (smaller or larger) representation of a physical phenomenon that can be manipulated independently of the real-life system it imitates. If, by using a model, a design fault in a reservoir can be shown to result in some catastrophe, then it follows that a similar event is likely to occur in the full-size reservoir. Model experiments are thus used to simulate and predict present and future behaviour patterns. Prediction of past behaviour is also often important, for instance, when parameter values must be synthetically reconstructed for times before observations were available. If the model is a scaled-down version of the reservoir (made of wood, steel, *papier maché*, etc.), then water can be introduced and observed and analogies drawn. Numerical modelling is more abstract: the reservoir is represented, not by a physical entity, but by a system of mathematical equations.

A Numerical Model

If any natural phenomenon can be represented by one or more algebraic equations, then these equations can be said to constitute a *mathematical model*. Since the approach is strictly quantitative, the model is a numerical one. A simple example may be useful here. If a reservoir is just overtopped and the water overflows into a bellmouth overflow, it may be necessary to have knowledge of the velocity of the water at the bottom of the bellmouth. A physical scale model could be built to find this out, but it would be costly and time consuming. However, it is known that all bodies that fall under gravity obey certain mathematical rules. One of these says that the final velocity, v, is equal to the square root of twice the acceleration due to gravity, g, multiplied by the distance, s (assuming that it starts to fall with zero velocity). The mathematical model is thus a single equation

$$v^2 = 2gs \qquad (10.1)$$

Putting numbers into this (that is, values for s) would give the corresponding values for v (see table 10.1). This could now be called a numerical model. Since this equation is simple enough to solve analytically (that is, using classical algebra), the nomenclature is perhaps a little inappropriate and the term 'numerical model' is usually reserved for those mathematical models that have no easy, analytic solution.

Table 10.1

s (m)	0	10	20	30	40	50
$2gs$	0	196	392	589	784	981
v (m/s)	0	14	20	24	28	31

Such sets of equations can be solved using paper and pen but rely on various approximation procedures; the study and development of such procedures forms a subject called *numerical analysis*. The complexity of the systems of equations solvable by these techniques is limited only by the calculating power and time available. Since, to a large extent, these are 'number-crunching' techniques, numerical models are highly suited to computer simulation.

Types of Model

There are essentially two types of model that are useful in hydrology: *stochastic* and *deterministic*. In a stochastic model, the values of the variables in the mathematical equations are only known with a certain probability of occurrence. A reservoir model in which the total outflows must be forecast as a function of the inflows may be stochastic if the time series of inflows used in the simulation is deduced by statistical techniques (for example, a flow of once in 100 years—see chapter 3). If all the variables can be regarded as free from any randomness, so that the variables can be calculated with certainty, then the model is deterministic.[1] Hindcasting of temperature profiles using observed meteorological parameters can be purely deterministic. Often the model type is less obvious or well defined. The predictive water quality model described earlier is assumed to be deterministic, since no parameters have a probability of less than one associated with them. However, some degree of randomness must exist in the meteorological forcing terms and if such statistical variations were taken into account the model would become stochastic.

Two further terms are sometimes used in the description of the model type: *conceptual* and *empirical*. The former term is applied to a physical approach based on a knowledge of the governing physical laws. An empirical model is one in which the analysis of observational data is used as the sole basis for the model formulation and no understanding of the physics is involved.

Finite Differences

Any model of a reservoir must include the currents at different depths, the temperature at different places, the biota and the other characteristic features discussed in previous chapters. Thus, different places across the lake and different depths may experience completely different physical and biochemical behaviour. To describe the lake perfectly, the complete set of equations must be solved separately at each position in the lake. This would necessitate an infinite number of calculations, since the lake water forms a *continuum*—a smoothly changing

---

[1] As the far-seeing reader will have realised, it follows from the above definitions that deterministic models are simply a subset of stochastic models. However, most researchers prefer to segregate the two types as discussed above.

mass whose properties (density, velocity, etc.) are different at neighbouring points. Even with a computer this number of equations is totally impossible and so the points at which the calculations are undertaken must be severely restricted.

To do this, the reservoir is divided into a discrete number of parts, both horizontally and vertically. If our concern is only with what happens at different depths (for example, in the formation of the thermocline—see chapter 7), then it may be assumed that the reservoir can be split into some number (say 25 to 100) of homogeneous horizontal slices, all of the same thickness (figure 10.1). In model-

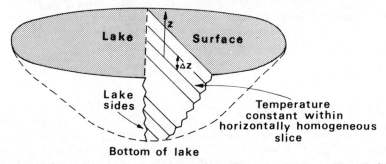

Figure 10.1 *Vertical cross-section through reservoir illustrating slice structure; each slice is horizontally homogeneous and of (vertical) thickness Δz*

ling the current at different places in the lake but at a constant depth, on the other hand, the appropriate approach would be to construct a two-dimensional model in the horizontal plane (at that one selected depth); a typical two-dimensional grid is shown in figure 10.2. In the case of the thermocline model, however, only one

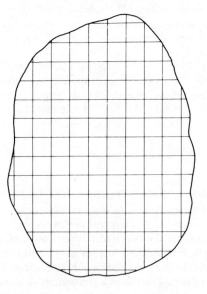

Figure 10.2 *Two-dimensional finite difference grid for modelling horizontal currents in a reservoir*

dimension (vertical) is being utilised. The value of any parameter (for example, temperature) is assumed to be constant for a given slice, irrespective of the horizontal, spatial location; this 'typical' value represents an average value over the depth of the slice (and across the horizontal areal extent of the lake). Figure 10.3

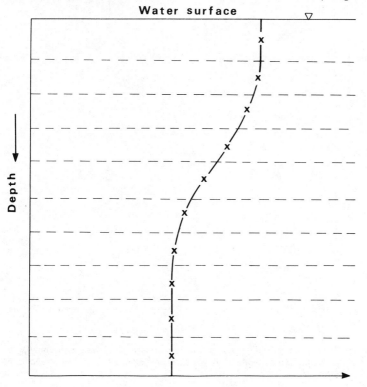

Figure 10.3 *Continuous temperature profile in a lake divided into ten slices of equal thickness; the temperature at the mid-point of each slice is indicated by a cross*

shows a typical vertical temperature profile. The dotted lines indicate the edges of the slices (in this case 10 in number) into which the reservoir is to be divided. If the mid-slice value of the temperature is assumed to be typical of that slice, then a numerical approximation to the temperature curve can be constructed (figure 10.4): a smooth curve has been replaced by a step curve. The final solution will thus be approximate, its accuracy depending on this representation.

The nonlinear equation for the calculation of the annual temperature changes within a lake (the heat transfer equation) can now be solved at each slice level. During this solution it is found necessary to calculate the temperature gradient, dT/dz, at each slice depth, z. If the temperature of the ith slice is T_i and the temperature of the $(i + 1)$th slice is T_{i+1} (see figure 10.5), then the temperature gradient between the two slices is the difference in temperature divided by the distance between the slices (that is, between the mid-points). For a slice thickness Δz, the temperature gradient is thus represented by

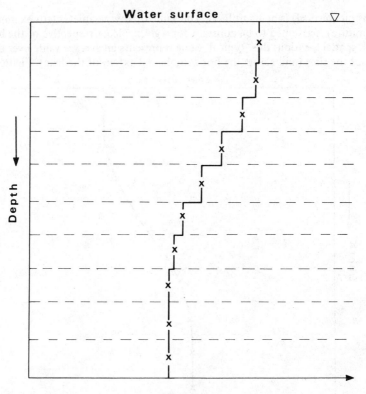

Figure 10.4 *Discrete, finite difference approximation to the temperature profile in figure 10.3; the mid-point temperature for each slice replaces the actual temperature at all locations within that slice*

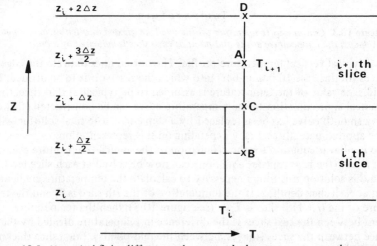

Figure 10.5 *Numerical finite difference scheme to calculate temperatures and temperature gradients*

$$\frac{dT}{dz} = \frac{T_{i+1} - T_i}{\Delta z} \tag{10.2}$$

This representation is known as a *finite difference* scheme and is a standard numerical technique for the approximation of a continuous function.

Strictly speaking, equation 10.2 gives the value of the temperature gradient at a point halfway between the mid-points, A and B, of the two slices—that is, at point C in figure 10.4. The equations must be solved at a consistent level usually deemed to be the level of the mid-points—that is, at points A and B, but not C. To obviate this problem one of two solutions is possible. In the first, the average temperature of the slices is calculated and assumed to be representative of point C

$$T_C = \frac{T_{i+1} + T_i}{2} \tag{10.3}$$

Similarly, the temperature at point D, T_D, is calculated as the average of the $(i + 1)$th slice and the ith slice. Now equation 10.2 is applied to the temperature gradient between points D and C to get the temperature gradient at the mid-point, which is now point A. Thus

$$\left(\frac{dT}{dz}\right)_A = \frac{T_D - T_C}{\Delta z} = \frac{(T_{i+2} + T_{i+1})/2 - (T_{i+1} + T_i)/2}{\Delta z}$$

$$= \frac{T_{i+2} - T_i}{2\Delta z} \tag{10.4}$$

Alternatively, if equation 10.2 is used directly to calculate the temperature gradients at points C and D, then the average of these two gives the value at point A, that is

$$\left(\frac{dT}{dz}\right)_C = \frac{T_{i+1} - T_i}{\Delta z}$$

$$\left(\frac{dT}{dz}\right)_D = \frac{T_{i+2} - T_{i+1}}{\Delta z}$$

Therefore

$$\left(\frac{dT}{dz}\right)_A = \frac{dT/dz_C + dT/dz_D}{2}$$

$$= \frac{(T_{i+1} - T_i)/\Delta z + (T_{i+2} - T_{i+1})/\Delta z}{2}$$

$$= \frac{T_{i+2} - T_i}{2\Delta z}$$

as before (equation 10.4).

An alternative numerical technique is the *finite element* method. Although the theory behind this is more complex and the time needed to program is longer, the computational time needed to solve the problem is shorter than for a finite difference scheme. Instead of representing the lake by a rectangular grid (see figure 10.2), a mesh of triangles or quadrilaterals is superimposed on the lake (figure 10.6). The parameter values are calculated as being constant across the area of each of the 'elements' and the resulting set of simultaneous equations is solved using a matrix method.

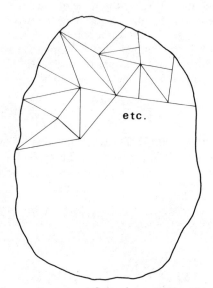

etc.

Figure 10.6 *Part of a two-dimensional finite element grid used to subdivide the horizontal surface of a lake*

Lake Simulations

To prove their worth, numerical models must be validated against real data. This can often prove difficult since the main forcing functions to the lake system are the atmospheric parameters (for example, sun and wind). These vary over such short time scales that it is not possible (with present-day computers) to simulate a year in the life of a lake, while taking into account all the changes in the weather. Meteorological observations are seldom available for the lake side and the weather 10 km away can be strikingly different at times. Windermere, for instance, is so large that this lake experiences a daily wind pattern (similar to the land–sea breeze phenomenon observed in coastal resorts) that is not observed 5 km away from the lake. Often, mean meteorological observations must be used. These do not include many of the sudden changes in weather (such as an intense and fast-moving depression) that are responsible for the most rapid changes in the lake structure. However, if data from many years (at least 30) are averaged, these transient phenomena disappear. These mean patterns give some indication of the long-run behaviour or *hydroclimate* of the lake (defined in a similar way to atmospheric climate).

Many numerical models (for example, those for thermocline formation) are ideally suited for this type of simulation since their formulation requires smoothed meteorological data. Figure 10.7 shows the hydroclimate predicted for Windermere using one such model. (Observed data for 1975 are shown in figure 10.8. Since 1975 was a hotter year than usual, the temperatures observed are higher than the average predicted temperatures shown in figure 10.7.)

In addition to the meteorology, the lake morphometry must be known and simulated. For example, figure 10.9 shows the variation of cross-sectional area with depth for the North Basin of Windermere.

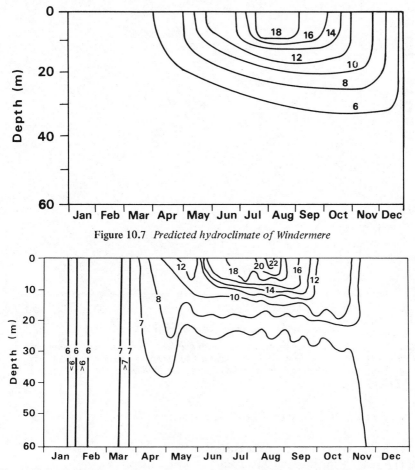

Figure 10.7 *Predicted hydroclimate of Windermere*

Figure 10.8 *Temperature structure for the North Basin of Windermere, 1975 (Isopleth values in °C.) (Courtesy of Freshwater Biological Association, unpublished data.)*

Programming

Any numerical model designed for use in a computer must be translated from its mathematical representation into a form suitable for input to a computer by

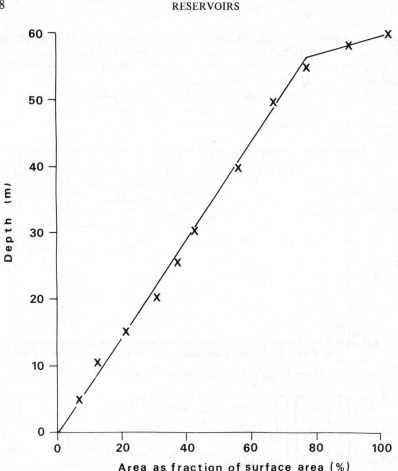

Figure 10.9 *Variation of cross-sectional area with depth for North Basin of Windermere; the crosses denote observations, while the curve is a linear approximation. [Source: A. E. Ramsbottom, 'Depth Charts of the Cumbrian Lakes', Sci. Publ. Freshwat. Biol. Ass., 33 (1976).]*

using a *programming language*.   here are four steps to writing any successful program

    (1) decide on the problem to be solved;
    (2) formulate the problem into a series of logical steps;
    (3) exemplify this logic by means of a flow diagram;
    (4) code the flow chart into a series of instructions written in the computer langauge to be used.

The first step may seem obvious, but the capacity of a computer to follow instructions slavishly may result in large volumes of output data. Sifting the desired results from the extraneous can be avoided if a little time is spent, at this first step, in deciding whether, for example, temperature, D.O. content or currents are required.

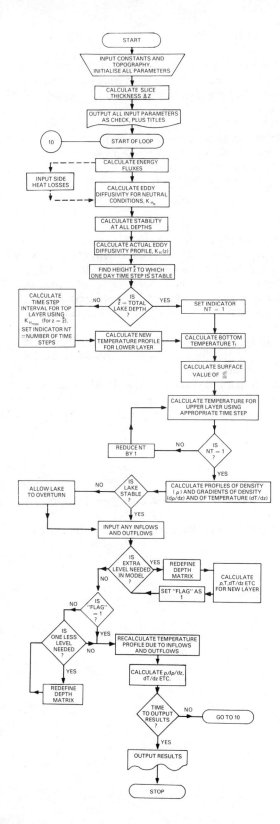

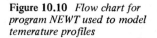

Figure 10.10 *Flow chart for program NEWT used to model temerature profiles*

The logic of the problem is bound up with the mathematics: the steps taken to a solution and the order in which these are undertaken must be appreciated. The logic is best illustrated in the form of a *flow chart* (see figure 10.10)—a tool useful both to oneself and to others who follow.

The final step of coding is mechanical. The programming language chosen will depend upon the machine available and its back-up—in the form of advisory services, computer library routines, etc. Most reservoir problems use Fortran or Algol and several detailed programming manuals exist.

Summary

If used with care, numerical models of reservoirs are a valuable tool in describing present events and also in predicting future behaviour. Simulations of currents, incoming water jets, temperatures and dissolved oxygen content provide useful information on the reactions that may be expected under a variety of conditions. (These conditions can be simulated—in either a numerical or a physical model—without waiting for the real-life conditions to occur; by that time it may be too late!)

Models employ several techniques found in numerical analysis. Since the solution is not analytic, it is only approximate although the error introduced can be controlled within the confines of the model.

Numerical modelling of reservoirs is a little over a decade old. Many of the advantages are yet to be realised, although real time implementation of some numerical models (especially systems models) are providing useful operating criteria for reservoir managers.

Suggested Reading

Smith, G. D., *Numerical Solution of Partial Differential Equations*, (Oxford University Press, 1969).

McCracken, D. D., *A Guide to FORTRAN IV Programming* (Wiley—Interscience, New York, 1972).

11

Flood Routing through Reservoirs

The prime objective of a reservoir is to collect and store water that is in excess of immediate requirements and then to make it available when there is a shortage. This shortage may occur in the vicinity of the reservoir or in another area of the country, reflecting the differing amounts of rainfall experienced by different regions.

A reservoir can also be used to lessen the effects of floods. As it has a large capacity, a reservoir can usually 'absorb' a flood. Only when a reservoir becomes full does excess water leave the reservoir to continue downstream. This *attenuating* effect of reservoirs on floods is well known but, for its maximum effect, the reservoir must be as empty as possible! This is in evident conflict with its water-supply function.

Flood Attenuation

If excess water is added to a reservoir then the level of water rises until it tops the dam, overspill weir or bellmouth. A reservoir that is not full has an unused storage capacity and thus can absorb and delay any flood waters that enter its upstream end. This storage capacity is proportional to the surface area, so that larger reservoirs have a greater attenuating effect. Even a full reservoir has an attenuating effect, since the flood peak is spread laterally over the total area before it reaches the downstream end of the reservoir. The peak is thus lowered and the discharge into the effluent stream is diminished. To calculate how effective a reservoir will be in this role of flood control, the concept of *continuity* is used.

Continuity states that 'what goes in must come out—or be stored in between.' Thus, the increase in the storage (written as ΔS) in a short time Δt is the difference between the volumetric rates of input, I, and output, Q. A simple mathematical equation expresses this as

$$\Delta S = (I - Q)\Delta t \qquad (11.1)$$

The output must of course allow for losses due to seepage and evaporation, but over a short period of time these can be neglected.

101

Hydrographs

The best pictorial method of representing a flow of water, either in a river or through a reservoir, is a *hydrograph* (see figure 11.1). A hydrograph is a continuous record of the volume of water (or *discharge*) passing a point as a function of time. It may be obtained at any point on a flowing river: for example, for the flow into or out of a reservoir—named the *input* and *output hydrographs* for the reservoir. This record is only produced if the flow is being monitored continuously, for example, by a pen recorder. If the flow is only measured at certain times of the day (manually usually) then a discrete set of measurements—say, every 3 or 6 hours—is obtained. It must then be assumed that the flow was constant at that measured value over the time period in question (in which case a plot of the results will be in the form of a *histogram*—see figure 11.2).

If the flow rate rises sufficiently it may approach 'flood' proportions. An idealised (or theoretical) *flood hydrograph* is shown in figure 11.3. The peak (shown at A) occurs as a very rapid response to heavy rainfall. Both the surface runoff and the groundwater component are seen to rise rapidly. The part of the flood due to surface runoff soon vanishes, but the flow due to the groundwater entering the stream (known as *base flow*) returns to its lower, dry-weather value

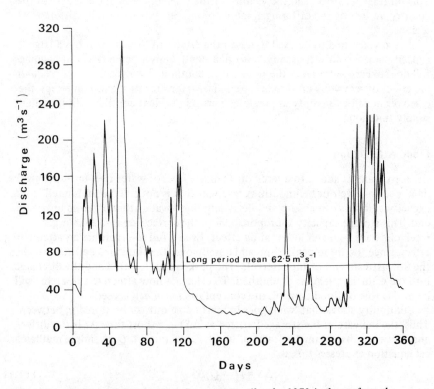

Figure 11.1 *Hydrograph for the River Severn at Bewdley for 1970 (redrawn from the Surface Water Year Book 1966--70, HMSO)*

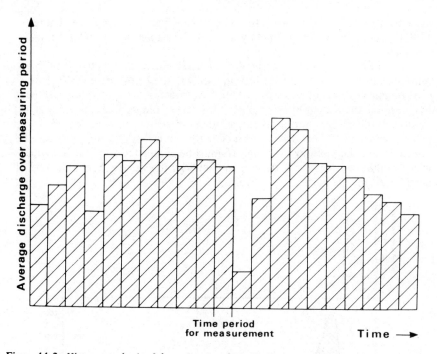

Figure 11.2 *Histogram obtained for a river in which the flow is only measured at a discrete number of time intervals*

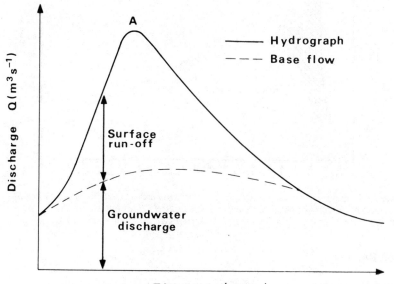

Figure 11.3 *Idealised flood hydrograph*

more slowly. The curve shown is smoother than a real hydrograph; see figure 11.4, in which the observed flood hydrograph for a single storm in the Pennines is depicted.

If the inflow rate can be measured and a hydrograph constructed, then the values of input I as a function of time are known; these are written as $I(t)$. The water engineer wants to know the attenuating effect of the reservoir and so needs to determine the output hydrograph $Q(t)$. To be of practical use, however, $Q(t)$ must be predicted and not measured, although measurement is sometimes a physical possibility. This is accomplished mathematically by using equation 11.1, but to do this the unknown variable ΔS must be eliminated from the equation, which requires knowledge of the relation between the storage, S, and the rate of outflow, Q, for the particular reservoir.

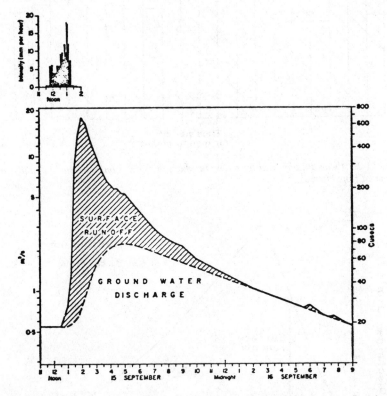

Figure 11.4 *A typical flood hydrograph resulting from a single rainstorm over a Pennine catchment in September 1961; drainage area 15.3 km² (5.9 sq. miles) and altitude of gauging station almost 305 m (1000 feet) O.D. (reproduced from Smith, 1972)*

Flood Routing

Nowadays most reservoirs overflow either over a weir or into a bellmouth over-flow (see figure 11.5). For a simple weir the amount of water flowing over it, Q, is related to the height, H, of the water above the top of the weir crest. This

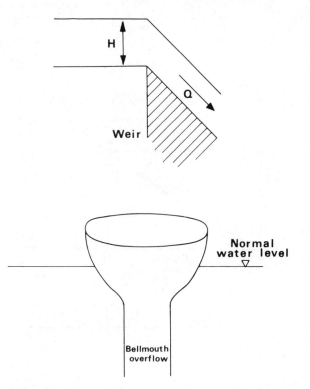

Figure 11.5 *Simple weir and bellmouth overflow—removal of excess water from a reservoir*

relationship may be expressible by a mathematical equation or may be derived from observational data. In both cases, a graph of the variation of Q as a function of S can be drawn; see figure 11.6 (upper graph). However, the storage, S, is represented by the volume of water in the reservoir and is thus a known function of the depth of water. Since the volume of water in the reservoir can easily be calculated for any given depth by using contour maps of the reservoir basin, a graph of S against H can be easily drawn. If this graph is constructed using the same scale for H, it becomes feasible to derive the storage for any given discharge (and vice versa). The lower graph of figure 11.6 shows this curve. Knowing the value of Q $(= Q_1)$, the value of H $(= H_1)$ may be deduced using curve 11.6a (indicated by point B). Moving vertically downwards on to curve 11.6b, the value of H is immediately transferred to this curve (point C) since the scales for H are identical. It is then possible to read off the value of S $(= S_1)$ corresponding to H_1 (see point D). Since S can now be found for a given value of Q, it becomes possible to use this graphical technique together with equation 11.1 to *route* the flood through the reservoir.

If the inflow to the reservoir is known, then the outflow can be calculated. To see what happens to the flood water throughout the duration of the storm, the output hydrograph must be constructed by performing the above calculation at each time step. Times of 0, 1, 2, 3, . . . units, each of say 3 or 6 hours, are chosen

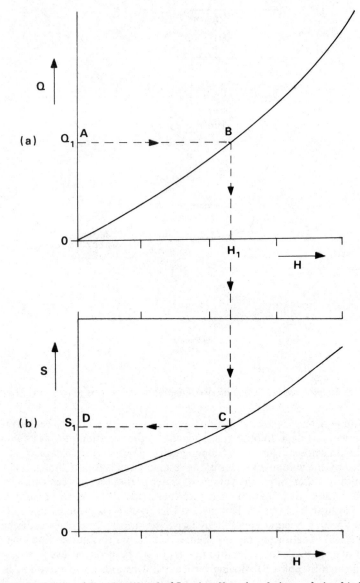

Figure 11.6 *Graphs of Q against H and of S against H used to derive a relationship between S and Q*

and the value of Q is found for the observed values of I as follows. The values of input flow I can be found directly from the measured input hydrograph at each time step (see figure 11.7). The following notation is used: $I_0 = I(t = 0)$, $I_1 = I(t = 1), I_2 = I(t = 2), \ldots$ Using the mathematical technique of *finite differencing* (see chapter 10), our equation is written

$$\frac{2S_{t+1}}{\Delta t} + Q_{t+1} = \frac{2S_t}{\Delta t} + I_t + I_{t+1} - Q_t \qquad (11.2)$$

where Q_t is the value of Q at any selected time t (that is, the present time step) and Q_{t+1} is the value at one time step further forward. If all the values are known initially and the input hydrograph is defined, then the right-hand side of this equation is already determined. Thus the left-hand side is also known and, using figure 11.6, Q_{t+1} can be calculated. This is then repeated at the next time step. The resulting hydrograph is shown in figure 11.8. (The calculation is a repetitive one and ideally suited for solution by computer or by using a calculator.)

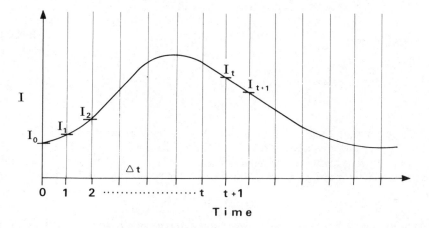

Figure 11.7 *Hydrograph divided into sections determined by the small time step Δt for use in a finite difference scheme; the continuous function $I(t)$ is thus converted into a sequence of inflow values I_0, I_1, I_2, etc. defined at discrete times $t = 0, 1, 2$, etc.*

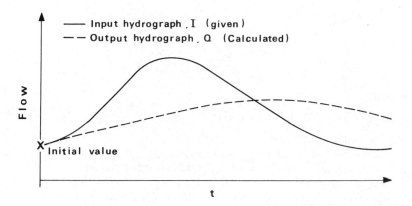

Figure 11.8 *Given input hydrograph and calculated output hydrograph for a flood routed through a reservoir*

The flood has thus been routed through the reservoir. Knowledge of how much the flood peak will be lowered will help the reservoir manager to decide whether flood warnings need to be issued downstream. As the water flows down the river further attenuation will occur as a result of the storage in the river channel, but that calculation is deemed outside the scope of this book.

Numerical Example

To illustrate this technique, and a slightly quicker method of solution, a numerical example is given.

Equation 11.2 can be rewritten as

$$Y_{t+1} = Y_t - 2Q_t + I_t + I_{t+1} \qquad (11.3)$$

where

$$Y_t = \frac{2S_t}{\Delta t} + Q_t$$

A single graph of Y against Q is more useful than the two graphs shown in figure 11.€ *but* can be derived from them.

For a reservoir with vertical sides and a surface area of 10^5 m^2, $S = 10^5 \times H$ m^3. For a simple weir of breadth B m, the relationship between Q and H is found to be

$$Q = 1.9BH^{3/2}$$

Thus for a weir 10 m wide and using a time step (Δt) of 6 h, the expression for Y becomes

$$Y = \frac{2S}{\Delta t} + Q = \frac{2 \times 10^5 H}{6 \times (60)^2} + Q$$

$$= \frac{2 \times 10^5}{6 \times (60)^2} \left(\frac{Q}{1.9B}\right)^{2/3} + Q$$

$$= \frac{2 \times 10^5}{6 \times (60)^2} \left(\frac{Q}{19}\right)^{2/3} + Q$$

Therefore

$$Y = 1.31Q^{2/3} + Q \qquad (11.4)$$

This is plotted in figure 11.9.

To calculate Q, we again start from knowledge of values for I_t, I_{t+1} and Q_t. A value for Y_t can then be found from figure 11.9. Thus all the terms on the right-hand side of equation 11.3 are known, and hence so is Y_{t+1}. Using figure 11.9 (the other way round), we can read off a value of Q_{t+1} from the value for Y_{t+1} that has just been found. This is the answer needed and thus the values of Q at all steps can be worked out.

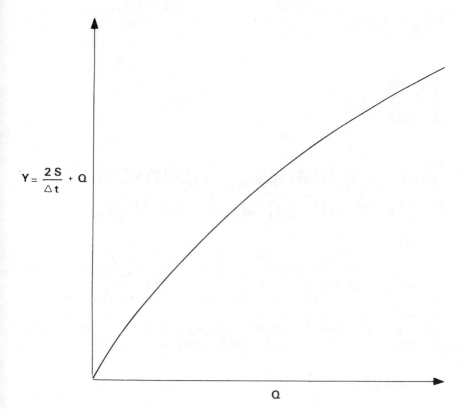

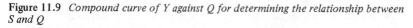

Figure 11.9 *Compound curve of Y against Q for determining the relationship between S and Q*

Summary

Although a reservoir is constructed in order to ensure a supply of water adequate to satisfy an increasing demand, it can often have a useful secondary function of flood control. If the reservoir is partially empty, then the flood waters entering the reservoir will be effectively slowed down, because of the large storage capacity available. Such an *attenuating* effect can be calculated for a reservoir of known size and contents. The process is described in terms of input and output *hydrographs*. The flood water is said to be *routed* through the reservoir. A simple *finite difference scheme* that allows this calculation to be performed numerically is introduced.

Suggested Reading

Smith, K., *Water in Britain* (Macmillan, London, 1972).
Wilson, E. M., *Engineering Hydrology*, 2nd ed. (Macmillan, London, 1974).

12

Water Quality, Treatment, Conservation and Re-use

The quality of water stored in a reservoir depends upon many things. The physical and chemical state, the type of biota present and the withdrawal point are all relevant in determining the quality of the water that will be abstracted for use. Seldom is this stored water immediately suitable for domestic use, and some treatment (usually chemical and biological) will be necessary. This may be as part of the reservoir installation or many kilometres away at a separate plant.

Water Quality

Water quality must be defined in terms of the usage required of the water. Industrial consumers may be able to accept and use water containing, for example, more metals and less oxygen than water used domestically. The appreciation of water quality and, correspondingly, the legislation that is necessary to maintain that quality are determined not only by the type of consumer but also by the prevailing 'climate of opinion' in society. The water quality that was acceptable in 1900 would not be accepted today: times and attitudes change. The expected state of health of a community can be related to the demands that it places on the quality of its domestic water supply.

Increasing expectations lead to pressure on the water supply engineer to exercise stricter control on both stored water quality and the treatments to which it is later subjected. Before the introduction of slow sand filters, cholera was endemic in many countries. In 1892, 8600 people died in Hamburg as a result of cholera, whereas the neighbouring town of Altona, although using the same water supply, suffered no casualties. Altona had installed slow sand filters. As a direct result, many more cities considered the introduction of sand filters and it was these improving standards in water quality and water treatment that resulted in the virtual extinction of the disease in Europe.

Pure water rarely occurs in Nature. The water that is closest to chemical purity is probably the large amount frozen in the Polar icecaps. Rain water, which is the primary input to reservoirs, either directly or else indirectly via stream flow and

Table 12.1 World Health Organization Drinking-Water Standards

Characteristic	Highest desirable level	Maximum permissible level
Total solids	500	1500
Colour (°H)	5	50
Taste	unobjectionable	–
Odour	unobjectionable	–
Turbidity (Formazin units)	5	25
Chloride	200	600
Iron	0.1	1
Manganese	0.05	0.5
Copper	0.05	1.5
Zinc	5	15
Calcium	75	200
Magnesium	30	150
Sulphate	200	400
Total hardness (as $CaCO_3$)	100	500
Nitrate (as NO_3)	45	–
Phenol	0.001	0.002
Anionic detergent	0.02	1.0
Fluoride	0.9–1.7 (mean temp. 12 °C) 0.6–0.8 (mean temp. 32 °C)	
pH (units)	7–8	min. 6.5 max. 9.2
Arsenic	–	0.05
Cadmium	–	0.01
Chromium (6+)	–	0.05
Cyanide	–	0.05
Lead	–	0.10
Mercury	–	0.001
Selenium	–	0.01
Polynuclear aromatic hydrocarbons	–	0.0002
Gross alpha radioactivity (pC/litre)	–	3
Gross beta radioactivity (pC/litre)	–	30

Concentrations are expressed in mg/l, except where noted.
The bacteriological standard for water in the distribution system is as follows.

(1) In 95% of samples examined throught a year coliform bacteria should be absent in 100 ml.
(2) No sample should contain *E. coli* in 100 ml.
(3) No sample should contain more than 10 coliform organisms per 100 ml.
(4) Coliform organisms should not be detectable in 100 ml of any two consecutive samples.

Source: *International Standards for Drinking-Water*, 3rd ed. (W.H.O., Geneva, 1971).

groundwater, contains not only carbon dioxide (absorbed during its descent through the atmosphere) but also small particulates—dust, salt sea spray, etc.— which have acted as condensation nuclei in the process of cloud formation. If the water enters storage by an indirect route (for example, via runoff), then, dependent upon the local geology, it will also have acquired oxygen, traces of carbonate ions, magnesium, calcium, decaying organic material (*detritus*) and an abundance of animal and plant life. Once in the reservoir, plant growths, animal reproduction and deaths (not to mention roosting seagulls) can add to the 'undesirable' load.

The over-all quality of the water is determined by its temperature and chemical state (for example, its pH, the amounts of each of the trace elements present, the oxygen levels), as reflected in the World Health Organization Standards for drinking-water in table 12.1. Although unspecified in these standards, the amount of dissolved oxygen often gives a good first approximation to the quality, especially if the stored water is known to be uncontaminated by, for example, heavy metals or other toxic substances. The amount of dissolved oxygen not only determines the quantity of life present in the water, but is also a desired attribute of potable water. (Unaerated water would taste 'flat' and insipid.) Insufficient oxygen levels may result in *septic* or *anaerobic* conditions in the lake, when noxious gases such as hydrogen sulphide (H_2S), methane (CH_4) and ammonia (NH_3) may be released.

Acidity is easily measured in terms of the pH—a measure of the hydrogen ions present. A value of 7 is assigned to neutral, pure water; larger values indicate alkaline conditions and lower values acidic conditions. It has recently been suggested that the possible role of the hardness of water in coronary diseases, about which there has been some speculation, may be of secondary importance and that the correlation with the acidity of the water is stronger.

Impurities in a water supply can be responsible for both short term epidemic outbreaks and long term disease caused, for example, by the accumulation of radioactive material. A high level of dissolved metals such as lead, cadmium and mercury represents a potential danger: contamination of a water body by any one of these may have fatal consequences. In 1953 a group of Japanese fishermen were all afflicted by a strange disease that was given the name Minamata Disease. The cause was finally discovered to be mercuric poisoning of the water and hence of the indigenous fish population, which formed the fishermen's staple diet.

Water Treatment

Britain nas long been a pioneer in water treatment. Treatment involves a combination of mechanical methods (filter beds) and chemical and biological techniques. Sand filter beds to remove suspended material were patented in Britain as far back as 1791, although the prototype did not come into operation until 1804 (at Paisley). Straining and microbiological techniques were developed and pressure filters were brought into use in 1870. Not until 1881 was chemical coagulation thought a useful process. Activated carbon has been in general use since 1920 as a purifying agent that acts as a taste remover. Other chemicals that have been, and in most cases still are, in general use are chlorine, ozone and fluoride. Chlorine has been used for sterilisation since 1910 (Reading) and is completely effective; excess dosings are usually removed by carefully adding calculated amounts of sulphur dioxide. Ozone has been used intermittently to sterilise water although its use today is limited to about seven water treatment works. Fluoridation remains a matter of strong controversy, and addition of the chemical, as a supposed check on caries (tooth decay), varies from Authority to Authority. The first trials were held as recently as 1955, in Anglesey.

Today, water treatment starts when the water is abstracted from the reservoir. To eliminate any algal growth, large biota and other debris, the water is screened and may pass through a coarse sand filter before it is pumped to the treatment works, some kilometres away (see figure 12.1). Here the water is filtered and *clarified*. Clarification is a further settlement process, catalysed by the addition of a chemical flocculant such as alum or polyelectrolytes. The sludge that forms is either scraped away or blown off by means of compressed air. The water is then chlorinated and passed on for any other treatment deemed necessary: fluoridation, phosphorus removal, softening, etc. It passes along covered channels into a service reservoir, which is completely enclosed and situated in close proximity to the demand area.

Figure 12.1 *Wing treatment works used to treat the supply from Rutland Water*

The treatment of sewage and industrial effluents, although it forms a part of the hydrological cycle, is too far removed from the subject of reservoirs to merit discussion here. Excellent texts on this subject are cited in the bibliography.

Water Conservation

Water conservation may be seen only as an extremely long term necessity or as a crisis measure. Yet with an increasingly fluctuating climate, if industrial and domestic conservation is applied *now*, the restrictions in use seen during the 1976 drought may be avoided.

Conservation is the minimisation of avoidable wastes, although it must be remembered that, in many supply systems of the world, the distribution pipe network is so old that leaks are numerous, possibly accounting for 25 to 40 per cent of all water supplied. Restrictive practices can be introduced in many forms.

On the domestic scene many conservation measures were advocated during the drought: bricks in cisterns, washing-up water for gardens, bathing with a friend. Many of the suggestions made were either fatuous or resulted in such hardship to the consumer as to be of only temporary use. However, such ideas embody many of the underlying problems. Domestic sanitation accounts for 35 to 40 per cent of all nonindustrial water usage. If no solids are to be flushed away, the standard 9-litre flush is unnecessary, and dual-flush cisterns are now regularly being installed in new houses. However, it has recently been shown that decreasing the volume of the flush within functional limits (by possibly 35 to 40 per cent) may create conditions more amenable for (possibly pathogenic) bacteria. Irrigation of the vegetable patch using waste water soon becomes a toil, but a gravity feed from bathrooms to storage and hence to the garden when desired could easily be installed at little cost. However, possibly the most efficient way to reduce usage in the home is to meter the supply! (see chapter 5), although the 1976 figures for domestic consumption derived by the Severn Trent Water Authority show little difference between the supplies at Malvern (metered) and Mansfield (unmetered). Socio-economic differences may be of importance in this context. Wash basins fitted with spray taps save about 67 per cent water, and atomiser taps can save 90 per cent of the water used for handwashing. The increasing use of showers in preference to conventional baths is beneficial in water saving, since five showers can be taken before the water of a single bath is used.

Industry can probably make even greater savings to a country's water bill. Once-through systems are wasteful. The Central Electricity Generating Board loses 0.38×10^6 m^3 of water per day by evaporation alone. (This figure rises to 0.67×10^6 m^3 when industrial cooling is included.) Cost-benefit analyses indicate that cooling for power generation cannot be improved. However industrial water usage can be restricted by using the sealed recirculatory systems now available. Not only does it save water, but it also saves the industry water charges and drainage charges.

Water Re-use

Water is already being continually re-used. River water abstracted near the estuary will probably contain treated effluent discharged into the water course well upstream. It is said that when you drink a glass of water in Central London, it has already been drunk by seven people before you.[1] More intensive cases of re-use abound. During a drought in Kansas, effluent from trickle filters was returned to storage reservoirs on a 20-day cycle. On a more permanent basis, one-third of the water supply of Windhoek in Namibia is derived from treated waste water and some toilets in Japan and America re-use water on a small semiclosed circuit.

The water need not be re-used for potable supply. It may be used to create water parks for recreation or in irrigation. Within industry it is possible to collect effluent from a treatment process needing high-quality water and re-use it for a second process requiring an inferior water quality. The industrial treatment applied to the water prior to re-use may be oxidation, chemical dosing, removal

[1] The same could be said of many of the world's largest rivers (for example, the Rhine).

of solids, ion exchange or reverse osmosis. In 1976 one factory in London manu-facturing semiconductor chips of silicon on a 24-hour basis introduced a recirculation technique which saved up to 9 m^3 (2000 gallons) an hour in a process needing only 11.25 m^3 (2506 gallons) an hour (plus 20 per cent to make up for unavoidable production losses). The costs of recycling are becoming more attractive to industry as more systems become available, although, if waste heat is among the by-products removed by the recirculating water, it must be remember-ed that discharging effluent at a much elevated temperature may create serious problems both to the physical environment and to the aquatic ecosystem.

Summary

The quality of water acceptable for human consumption has a legal definition. Improving technology implies that water quality standards will become progres-sively more stringent. Water may be unacceptable because of the high amount of dissolved or suspended solids, a high (or low) pH or insufficient oxygen, or because it contains harmful organisms or chemicals. The quality is improved by a variety of treatment methods, often specific to a single source. Sources in Great Britain, however, are disinfected (usually by chlorination) as a health precaution.

Conservation can be undertaken both by industrial and domestic consumers, although this may introduce an unnecessary health hazard or a final discharge at a temperature that is undesirably high for the aquatic ecosystem. Water re-use is presently carried out on various scales. Treated effluent may be used for irriga-tion, recreation or even as a potable source. This depends upon the availability of other resources. Thus, at present, arid countries may find it more acceptable than areas with dependable rains. Better housekeeping and an acceptance of treated waste water for reconsumption may be vital in years ahead if the water industry in all parts of the world is to continue to meet the ever-increasing demands made upon it.

Suggested Reading

Tebbutt, T. H. Y., *Water Science and Technology* (Murray, London, 1973).
Tebbutt, T. H. Y., *Principles of Water Quality Control*, 2nd ed. (Pergamon, Oxford, 1977).

Bibliography

Books on Sewage Treatment

Bolton, R. L., and Klein, L., *Sewage Treatment. Basic Principles and Trends* (Butterworth, London, 1971).

Fair, G. M., Geyer, J. C., and Okun, D. A., *Water and Wastewater Engineering*, Vol. 2 (Wiley, New York, 1968).

Hammer, M. J., *Water and Wastewater Technology*, (Wiley, New York, 1975).

Tebbutt, T. H. Y., *Water Science and Technology*, (Murray, London, 1973).

Tebbutt, T. H. Y., *Principles of Water Quality Control*, 2nd ed. (Pergamon, Oxford, 1977).

White, J. B., *Wastewater Engineering*, 2nd ed. (Edward Arnold, London, 1978).

Other Books[1]

Arthurs, A. M., *Probability Theory*, (Routledge & Kegan Paul, London, 1965) [3].

Belcher, H., and Swale, E., *A Beginner's Guide to Freshwater Algae* (H.M.S.O., London, 1976) [9].

Bierman, H., and Smidt, S., *The Capital Budgeting Decision* (Macmillan, New York, 1960) [5].

Fair, G. M., Geyer, J. C., and Okun, D. A., *Water and Wastewater Engineering*, Vol. 2 (Wiley, New York, 1968) [8, 9].

Golterman, H. L., *Developments in Water Science 2. Physical Limnology* (Elsevier, Amsterdam, 1975) [7, 8, 9].

Hammer, M. J., *Water and Wastewater Technology* (Wiley, New York, 1975) [6, 8, 9].

Hutchinson, G. E., *A Treatise on Limnology*, 3 Vols (Wiley, New York, 1967) [7, 8, 9].

Institution of Water Engineers and Scientists, *Manual of British Water Engineering Practice*, ed. W. O. Skeat, 4th ed. (Heffer, Cambridge, 1969) [1-6, 12].

---

[1] Numbers in square brackets indicate relevant chapters in this book.

James, E., O'Brien, F., and Whitehead, P., *A Fortran Programming Course* (Prentice-Hall, London, 1970) [10].

Maass, A., and Anderson, R. L., *And the Desert Shall Rejoice: Conflict, Growth and Justice in Arid Environments* (MIT Press, Cambridge, Mass., 1978) [2].

McCracken, D. D., *A Guide to ALGOL Programming* (Wiley-Interscience, New York, 1962) [10].

McCracken, D. D., *A Guide to FORTRAN IV Programming* (Wiley-Interscience, New York, 1972) [10].

Mills, D. H., *An Introduction to Freshwater Ecology* (Oliver & Boyd, Edinburgh, 1972) [9].

Okun, D. A., *Regionalisation of Water Management. A Revolution in England and Wales* (Applied Science, London, 1977) [2].

Overman, M., *Water* (Aldus, London, 1968) [1, 2].

Reid, G. K., and Wood, R. D., *Ecology of Inland Waters and Estuaries* (Van Nostrand, New York, 1976) [7, 8, 9].

Singh, J., *Operations Research* (Penguin, Harmondsworth, 1968) [6].

Smith, G. D., *Numerical Solution of Partial Differential Equations* (Oxford University Press, 1969) [10].

Smith, K., *Water in Britain* (Macmillan, London, 1972) [1, 11].

Smith, N., *A History of Dams* (Peter Davies, London, 1971) [2].

Smith, N., *Man and Water* (Peter Davies, London, 1976) [1, 2, 12].

Swinscow, T. D. V., *Statistics at Square One* (British Medical Association, London, 1976) [3].

Tebbutt, T. H. Y., *Water Science and Technology* (Murray, London, 1973) [12].

Tebbutt, T. H. Y., *Principles of Water Quality Control*, 2nd ed. (Pergamon, Oxford, 1977) [8, 9, 12].

Thomas, H. H., *The Engineering of Large Dams*, 2 vols. (Wiley, New York, 1976) [4].

Twort, A. C., Hoather, R. C., and Law, F. M., *Water Supply*, 2nd ed. (Edward Arnold, London, 1974) [1, 3, 4, 8, 9].

Walker, R., *Water Supply, Treatment and Distribution* (Prentice-Hall, Englewood Cliffs, N.J., 1978) [1, 6, 8, 9, 12].

Walsh, G. R., *An Introduction to Linear Programming* (Holt, Rinehart & Winston, London, 1971) [6].

Wetzel, R. G., *Limnology* (Saunders, Philadelphia, 1975) [7, 8, 9].

Wilson, E. M., *Engineering Hydrology* 2nd ed. (Macmillan, London, 1974) [3, 11].

Papers

Arden, T. V., 'Soft Water and Heart Disease', *Wat. Serv.* 80 (1976) pp. 250-2 [12].

Beaumont, P., 'Man's Impact on River Systems: A Worldwide View', *Area*, 10 (1978) pp. 38-41 [2].

Davis, R. K., and Hanke, S. H., 'Conventional and Unconventional Alternatives for Water Supply Management', *Wat. Resour. Res.*, 9 (1973) pp. 861-70 [5].

Fast, A. W., Lorenzen, M. W., and Glenn, J. H., 'Comparative Study with Costs of Hypolimnetic Aeration', *Proc. A.S.C.E., J. Envir. Engng. Div.*, 102 (1976) pp. 1175-87 [8].

Heaney, S. I., 'Some Observations on the Use of the *in vivo* Fluorescence Technique to Determine Chlorophyll-*a* in Natural Populations and Cultures of Freshwater Phytoplankton', *Freshwat. Biol.*, 8 (1978) pp. 115-26 [8].

Henderson-Sellers, B., 'Role of Eddy Diffusivity in Thermocline Formation', *Proc. A.S.C.E., J. Envir. Engng. Div.*, 102 (1976) pp. 517-31 [7].

Henderson-Sellers, B., 'Forced Plumes in a Stratified Reservoir' *Proc. A.S.C.E., J. Hydrol. Div.*, 104 (1978) pp. 487-501 [7, 8].

Henderson-Sellers, B., 'The Longterm Thermal Behaviour of a Freshwater Lake', *Proc. I.C.E.* 65 (Part 2), Technical Note No. 196 (1978) pp. 921-7 [10].

Herrington, P., 'The Economics of Water Supply and Demand', *J. Econ. Ass.*, 12 (1976) pp. 67-84 [5].

Middleton, R. N., Saunders, R. J., and Warford, J. J., 'The Costs and Benefits of Water Metering', *J. Inst. Wat. Engng and Scient.*, 32 (1978) pp. 111-22 [5].

Octavio, K. A. H., Jirka, G. H., and Harleman, D. R. F., 'Vertical Heat Transport Mechanisms in Lakes and Reservoirs', *M.I.T. Tech. Rep. 227* (1977) [7].

Ramsbottom, A. E., 'Depth Charts of the Cumbrian Lakes', *Scient. Publs Freshwat. biol. Ass.*, 33 (1976) [10].

Thackray, J. E., 'Metering Water Demand', *Water*, 19 (1978) pp. 7-9 [5].

Thackray, J. E., Cocker, V., and Archibald, G., 'The Malvern and Mansfield Studies of Domestic Water Usage', *Proc. Inst. civ. Engrs*, 68, part 1 (1978) pp. 37-61 [5].

Youngman, R. E., 'Observations on Farmoor: A Eutrophic Reservoir in the Upper Thames Valley, during 1965-1973', Paper No. 7, *Water Research Centre Symposium on The Effects of Storage on Water Quality* (24-6 March, 1975) [8].

Glossary of Terms

aerobic	a process utilising oxygen; or a condition in which the reservoir contains free dissolved oxygen
algae	single-celled photosynthetic plants (chlorophyll-bearing organisms)
anaerobic	a process that can occur in the absence of oxygen; or a condition in which the reservoir is devoid of free dissolved oxygen
aquifer	a stratum of rock that can hold water
autotroph	an organism that is able to gain energy directly from solar radiation by photosynthesis
biota	the animal and plant life in an ecosystem
ciliate	an organism belonging to the protozoan subphylum *Ciliophora* and possessing cilia (hairs)
consumption	the amount of water used
cost-benefit analysis	an economic approach in which total benefits (real plus shadow) are compared with anticipated costs
cumec	cubic metre per second
cusec	cubic foot per second
decomposer	an organism (for example, a bacterium or a fungus) that breaks down dead organic matter to produce energy for itself
deterministic model	a representation of a real system in which all parameter values are known with certainty
diel	relating to a 24-hour cycle
diffusion	a process in which random motions produce dispersion of the property under consideration
dipole	a molecule in which there is a separation of electrical charge
discharge	the volumetric flow rate
discounting	a procedure in economics to compare the costs and benefits achieved at different times over a long period

dissolved oxygen (D.O.)	the concentration of oxygen that is dissolved in water and is thus available for biochemical activity (kg/m^3, mg/litre, p.p.m.)
drawoff tower	a tower containing several withdrawal valves at different depths in the reservoir
dry weather flow (D.W.F.)	the flow in a river that is the result of a long period of low rainfall; the term has several differing statistical definitions
ecosystem	a functional system that includes the organisms from a natural community together with their environment
epilimnion	the warmer upper layer of a stratified reservoir
eutrophication	the biological process whereby an aquatic environment sustains an increased biological production as a result of enrichment with plant nutrients
failure	a reservoir system can fail in two distinct ways: failure of a reservoir dam implies fracture of the dam itself and accidental release of stored water. Failure of the reservoir itself occurs when the reservoir cannot meet the demands made on it and may possibly run dry
fetch	the length of open water surface over which the wind blows; longer fetches permit the wind to generate higher waves
finite difference scheme	a mathematical method by which a continuous derivative is approximated by a set of discrete values, thus converting a differential equation into a form suitable for computer solution
flagellate	an organism that propels itself by means of a flagellum (a whip-like protrusion)
groundwater	water present in subsurface rock strata
hardness	a chemical property of water associated with the presence of magnesium and calcium salts
heterotroph	an organism that feeds on other plants and/or animals— the opposite to autotroph
hypolimnion	the lower cooler layer of a stratified reservoir
inelastic or price- inelastic	a demand that exhibits no change as a result of a change in price is said to be inelastic
isopleth	a line joining all places with the same value of the variable in question
lentic	slow-moving (as of lakes); cf. lotic (as of rivers)
limnetic zone	an area of lake too deep for plants to root
littoral zone	the lake-side area, in which rooted plants thrive
loading	the amount of, for example, nutrient input to a reservoir, often expressed as weight per square metre of lake surface
minimum acceptable flow (M.A.F.)	the lowest flow to ensure the maintenance of water quality in a river
metalimnion	the middle layer of a stratified lake, in which the temperature gradient is greatest
mgd	million gallons per day
morphometry	measurement of the shape of a lake basin

nutrient	a chemical element that is used by living cells for sustaining life functions
oxidation	a chemical reaction in which a compound or radical loses electrons—often a reaction between oxygen and another element
pathogen	a micro-organism that is responsible for disease
pH scale	a measure of acidity (high values are alkaline, a value of 7 is neutral), defined as the logarithm of the reciprocal of the hydrogen ion concentration (measured in moles per litre)
photosynthesis	a process utilised by plants to produce carbohydrates from atmospheric carbon dioxide, a hydrogen source (such as water) and sunlight
Planck curve (black-body curve)	photosynthetic plankton a curve of radiation intensity as a function of wavelength
plankton	aquatic organisms that drift more or less passively with the water (cf. *nekton*, which are free-swimming aquatic animals, essentially independent of water movement, i.e. self-propelled)
potable	of consumable quality for humans (drinkable)
residence time	the average time for water to remain in a reservoir (*retention time*)
respiration	the release of energy and carbon dioxide from reduced carbon compounds (lipids, carbohydrates, proteins, etc.)
reverse osmosis	a method of water treatment used in desalination in which water is forced under pressure through a semi-permeable membrane which retains most of the salt
riprap	stone used to cover upstream faces of earth dams
salts	the products (together with water) obtained in neutralising bases with acids
saturation	the state in which the maximum amount of a solute (for example, oxygen) is dissolved in the solvent
scour	the rapid emptying and cleaning of a reservoir
shadow costs and benefits	costs and benefits that cannot be included in capital costs and benefits (for example, aesthetic beauty)
shuttering	a mechanism in the dam wall (of concrete dams) through which large volumes of water can be released
slip circle analysis	a method for determining whether an earth dam design will be stable under any conditions. Unbalanced moments are computed within several slip circles and these moments must be resisted by the strength of the dam material itself to ensure stability
stochastic model	a model in which parameter values possess some degree of uncertainty
stratification	the formation of a nonlinear temperature gradient and of a layered structure in a reservoir
supersaturated	containing an amount of solute greater than that at saturation (the theoretical maximum)

telemetry	a system in which data are recorded and transmitted by electromagnetic means and received and interpreted at a remote point
thermocline	the depth at which the temperature gradient is greatest in a stratified lake
topography	the description of the vertical variations in the bottom shape of a reservoir
trophic level	the relative position of a plant or animal in the food chain
turbidity	a reduction in the capacity of water to transmit light, resulting from the presence of suspended material
top water level (T.W.L.)	the maximum level of water permitted in a reservoir; any excess water entering the reservoir is allowed to overflow
zooplankton	nonphotosynthetic plankton (animals)

Conversions

1 m	= 3.281 ft
25.4 mm = 0.0254 m	= 1 in.
1 km	= 0.621 mile
1 μm	= 10^{-6} m
1 hectare = 10^4 m^2	= 2.471 acres
1000 kg	= 0.9842 ton
0.001 kg/m^3	= 1 mg/litre
1 m^3	= 219.3 gallons
1 cumec = 1 m^3/s	= 33.315 cusec = 19.005 m.g.d.
1 m^3/day	= 1000 litre/day
Temperature in K	= Temperature in $^\circ$C + 273.16

Index